# Brain
## vs
# Computer

**The Challenge of the Century is Now Launched**

**Second Edition**

## Other World Scientific Titles by the Author

*Near Field Optics and Nanoscopy*
ISBN: 978-981-02-2349-6

*Is Man to Survive Science?*
ISBN: 978-981-4644-40-2
ISBN: 978-981-4644-41-9 (pbk)

*Brain vs Computer: The Challenge of the Century*
ISBN: 978-981-3145-54-2
ISBN: 978-981-3145-55-9 (pbk)

*Longevity in the 2.0 World: Would Centenarians Become Commonplace?*
ISBN: 978-981-120-116-5
ISBN: 978-981-120-203-2 (pbk)

*Transhumanism: A Realistic Future?*
ISBN: 978-981-121-138-6
ISBN: 978-981-121-210-9 (pbk)

*Brain vs Computer: The Challenge of the Century is Now Launched Second Edition*
ISBN: 978-981-122-500-0
ISBN: 978-981-122-626-7 (pbk)

# Brain
## vs
# Computer

**The Challenge of the Century is Now Launched**

**Second Edition**

Jean-Pierre Fillard

**World Scientific**

NEW JERSEY · LONDON · SINGAPORE · BEIJING · SHANGHAI · HONG KONG · TAIPEI · CHENNAI · TOKYO

*Published by*

World Scientific Publishing Co. Pte. Ltd.

5 Toh Tuck Link, Singapore 596224

*USA office:* 27 Warren Street, Suite 401-402, Hackensack, NJ 07601

*UK office:* 57 Shelton Street, Covent Garden, London WC2H 9HE

**Library of Congress Cataloging-in-Publication Data**

Names: Fillard, J. P., author.

Title: Brain vs computer : the challenge of the century is now launched / Jean-Pierre Fillard.

Description: Second edition. | Singapore ; Hackensack, NJ : World Scientific
    Publishing Co. Pte. Ltd., [2021]

Identifiers: LCCN 2020034550 | ISBN 9789811225000 (hardcover) |
    ISBN 9789811226267 (paperback) | ISBN 9789811225017 (ebook) |
    ISBN 9789811225024 (ebook other)

Subjects: LCSH: Neuromorphics. | Artificial intelligence. |
    Human-computer interaction. | Cognitive science.

Classification: LCC TA164.4 .F55 2021 | DDC 006.3/82--dc23

LC record available at https://lccn.loc.gov/2020034550

**British Library Cataloguing-in-Publication Data**

A catalogue record for this book is available from the British Library.

For any available supplementary material, please visit
https://www.worldscientific.com/worldscibooks/10.1142/11957#t=suppl

Typeset by Stallion Press
Email: enquiries@stallionpress.com

"There is a flaw in that plan!" squeaked Scronkfinkel

— Nick Bostrom
*Superintelligence*

# Prologue

Why a second edition? There are two possible answers: as follow-up to a bestselling first edition, and that many things have changed in the meantime.

So, some words would be required to introduce this second edition. Keeping up with the news of our times is a very fitting and challenging task necessitating an update. This is the aim of this second edition which is a fully revised 2020 version of *Brain vs Computer: The Challenge of the Century* published in 2016.

When WSPC approached me to write a second edition, I jumped onto the Internet to discover the latest developments in the domain of Artificial Intelligence (AI) and was stunned by the evolution of the field in so short a while.

Thanks to John Seabrook,[1] I discovered that AI was ready to "percolate" through our minds with "Smart Compose", a feature introduced by Google in 2018, promptly followed by a "supercharged version": GPT-2.[2] As Greg Brockman cautiously puts it: "These machines haven't been taught to 'love humanity'"!

What, then, is this all about? Today when you enter a text in a computer, the machine will try to understand what you intend to write and

---

[1] "The next word: Where will predictive text take us?", John Seabrook, *New Yorker*, October 14, 2019. Available at: https://www.newyorker.com/magazine/2019/10/14/can-a-machine-learn-to-write-for-the-new-yorker#

[2] The release of that software was delayed, it seems, because it was considered "too good at writing"!

make corresponding speculations to the point that it supersedes the writer to complete his text with its own version.

This is the first example of a clear attempt of a machine to fully compete with a human brain and replace it over time.

So I am afraid that, in a near future, such a super-intelligent computer could propose to WSPC a third version of my book and become the celebrated author!

JP Fillard

*Author's Note:* Before sending the manuscript to the editor I would like to add a new comment about the recent pandemic ravaging the whole planet. This tiny virus reminds us that Nature has not been tamed and it is always there to impose some humility in our triumphant walk towards "progress".

We are taken back to the Middle Ages when there were no remedies to the various plagues. All our accumulated knowledge, all our intelligence (artificial or not), all our instrumentation are (at the moment) of no use to get a viable solution to this biological scourge. Our sophisticated microscopes send us pretty images of the culprit but no way to get rid of it!

Some transhumanists would propose to mutate us into better performing electronic android copies and that would be able to protect us from biological threats ... but, alas, computer viruses will still be there!

# Contents

# Preface

This is a second edition. As a consequence deep changes have been introduced to take into account the most recent evolution in the field. The first part devoted to brain specifications has been kept almost the same because the human brain has not changed much in the past few years! Part 2 includes important new considerations whereas Part 3 has also been modified.

Books, for centuries, were repositories of existing knowledge. They were carefully written and printed (when they were not handmade!). The knowledges of the times were considered as relevant for many years as the things evolved rather slowly. So any new book required making reference to the previous ones in order to help the reader to properly cover the subject.

With the advent of the 2.0 facilities, things have drastically changed. The timeliness of the culture evolves so fast that writing a book no longer allows the author to follow the speed things evolve. However the book is still the best haven to adopt a global approach, stand back and take a broad perspective. To do that and nevertheless be be kept abreast of the news, it has become necessary to make references to papers on the internet. Such references are rather volatile, so how else to go about it? The other way to put things right is to go for a new edition; that I did!

"To be or not to be a human?" This is the Shakespearean question the robot suggests on the cover of that book.

Nowadays the computer is undoubtedly becoming more and more "intelligent". This mineral object is doomed to soon constitute an intermediate state between the unanimated world and the living one. To what extent? This is the topic of the book.

It is now well known that every living species faithfully obeys Darwin's law of improved adaptation to the environment. To be precise, every species except human beings. This particular species was given a special divine gift: intelligence.

This "property," so particularly difficult to establish, immediately allowed Man to oppose the Darwin predications: the way now will be to adapt the environment for the comfort of man. This was done by first inventing adapted tools and then machines, which immediately contributed in an indisputable progress in living and working conditions.

Intelligence grew progressively with a better knowledge of our environment and our own nature. This was a cumulative process which benefited everybody, even if not everybody contributed. Intelligence applies to any domain of the human brain's activity, not only to mathematics!

Machines gradually and efficiently replaced humans in the tasks they were not able to fulfill themselves. Currently, humans get tired and dissatisfied with being intelligent — why not assign to machines the role of being intelligent instead of humans? They would surely perform better and take care of us still better.

This is the challenge we are dealing with in this book which does not require being a scientific specialist to be read. This is just a (pleasant) survey of a topic which we are all involved in. Proposed references often lead to websites or famous magazines; I am sorry they could be volatile and quickly obsolete but that is the price to pay for sticking to the most recent events. So many transformations are to appear in a near future that it is worth to be informed of what could happen.

Hope you will enjoy this book!

# Introduction

Since the turn of the 20th century, science and technology have been steadily bringing us new elements of knowledge, new investigative instruments of increasingly good quality, and new drastic changes in our daily way of life. The acceleration of the evolution is clearly evident. The global research community is in turmoil, taking up more and more human and funding resources.[1]

All of this has engendered metaphysical questioning the likes of which until recently belonged to fiction science but has since increasingly become part of the reality we are facing. Computers have filtered into our everyday lives and their dazzling and intrusive progress raises the worrying question of their inescapable cohabitation with humans.

Gone are the days of the computer being referred to as "a stupid machine only able to perform arithmetic" (certainly faster than us). Since these not-so-ancient times it has become evident that Artificial Intelligence (AI) will overrun our daily lives. The man–machine rivalry has become a recurring theme in view of the recent and substantial progress of AI.

A prophetic sign[2] was given by the European Parliament which had to debate a motion about the status of "Electronic Persons" that might be

---

[1] *Is Man to Survive Science?*, Jean-Pierre Fillard, World Scientific, 2015.
[2] "Les robots, des 'personnes electroniques', selon une motion enc ours d'examen au Parlement europeen", Reinformation.tv, June 22, 2016. Available at: http://reinformation. tv/robots-personnes-electroniques-motion-parlement-europeen-dolhein-56949-2/

given to robots if they extensively replace humans in some previously exclusive areas.

## Pandora's Box

The performances of computers are such that, in some specific domains, they outperform, without mercy, the human brain. How far can we go too far in such a man–machine competition? This question has been raised by the philosopher Nick Bostrom in his bestseller.[3] Following that, some distinguished minds, and far from the only, minds expressed their agreement with this theory and also their particular concern: among others, the physicist Stephen Hawking; Bill Gates, the former boss of Microsoft; and Elon Musk, bold creator of Twitter. All of them voiced their worries in light of the rapid advances in AI and its implementation in society.

One may also be note this unexpected (?) coincidence: we began to emphasize a detailed physical analysis of our brains at the very moment when the computer showed up — the competition was engaged right away!

The subconscious, which forms the bulk of our cerebral activity, is still very hard to access; for the brain, rationality requires an effort to concentrate whereas for irrationality the transcendence is spontaneous, originating from who knows where in the depths of the mind.

There is the very place for what we call instinct, intuition, feeling, sentiments, individual personality, the innate sense of good or bad, beautiful or ugly! Everything which does not belong to any kind of causality. As an example of that field, we will detail the domain of the arts and more especially the art of painting. Such an activity does not refer to a particular necessity or rationality with respect to our needs for the living but is nevertheless largely appreciated.

Our brain, this black box so far physically inaccessible, is beginning to yield some of its intimate secrets, well-hidden up to now. Our means of

---

[3] Nick Bostrom is a professor at Oxford University, where he heads the Institute for the Future of Humanity. In 2014, he published *Superintelligence: Paths, Dangers, Strategies*, Oxford University Press.

external investigations, such as CAT,[4] MRI,[5] EEG,[6] or PET scan,[7] provide more detailed information but nevertheless only give us images of the indirect consequences of our cerebral activity; the basic mechanism, still remains to be revealed, along with the biological and genetic functioning of the neuron. Cognitics is a very recent science which is still finding its way but move forward very fast.

We are here opening the Pandora's Box and God only knows what could emerge out of it (good or dreadful?). Does this means that the computer can in the medium term compete with, complement, or even profitably replace the human brain in a mock "transposed life"? The question is now pending. Some androids already exist which behave with a disturbing realism, even though they remain unable to "think." Other machines, like Watson at IBM, get trained with increasing success to "understand."

The day when (and the time will come) we know how to modify a biological brain or how to create an artificial brain, what will be achieved? Geniuses, valuable disabled people or "improved" ordinary people? Improved in what? Reinforced memory, Internet-connected culture, special skill in mathematics or music? Who to be taken as a model, and who could have the merit of being reproduced and perpetuated?

In any case, knowing if it is feasible to mimic a brain with a computer is, as of now, a spurious question. We cannot imitate something we don't know and, also as of now, we are not about to solve every mystery of the mind even if, undeniably, the knowledge is keeping pace. Moreover, the day we will know everything of the brain, many other ethical questions will surely emerge which we do not yet have a clear idea of. But one thing is for sure: the scientists now talk freely, without feeling embarrassed about these questions and without being accused of going into science fiction.

---

[4] Computer Assisted Tomography, the new versions of which allow a sharp imaging of the inner body with a very reduced X-Ray exposure.

[5] MRI: Magnetic Resonance Imaging which now operates in real time and is called "functional."

[6] EEG: Electro Encephalo Gram. An "old" method for recording evoked electrical potentials which is still valid and benefits from continuous improvements in signal analysis.

[7] Positon Emission Tomography, a smart imaging technique requiring radioactive products.

Obviously, however, the computer, once considered as "a stupid machine," continues to progress toward a fantastic storage capacity and an increasingly subtle versatility. It imposes itself on our daily lives and even our mores in an increasingly pressing way. It rises from any direction.

In this new story what happens to our self, our soul, our personality so dear to the philosophers and the religious? Are we biological beings, purely material, changing, and transitory; are we able to be copied in a vulgar machine as easily as we can take a snapshot? Is everything to be explained, and are we lacking immanence? The knowledge we are gathering on our neuronal intimacy will no doubt affect our religious views. Where is God in this century's challenge?

Nevertheless, the questions that one may already pose abound and will have to be addressed quickly. We will try in this book to focus on some causes for reflection on these burning issues, limited to the present state of our knowledge; but everything can be thrown into question as to the pace of the new instruments and the new discoveries.

Some see the computer as a means to enhance our intellectual performances while remaining masters of the game; others see in it an infernal toboggan leading to total dependence on mechanical rationality. Each specialist preaches to his parish, and both sides already fight, each pitching their own arguments.

Distinguished minds already send out an alarm on the rising performances of AI and the threats they could pose to the "Natural Intelligence", so unequally distributed in our human species but which has still allowed us to take a few more steps forward until the primitive ages.

All of this goes back to my great-great-grandfather more than a century ago; he was a country man and saw for the first time a steam-powered engine running on a railway. He was, of course, all afraid, and he screamed: "Nothing is pulling in front, nothing is pushing behind and yet it moves!" Unthinkable in those days! Today, what could be unthinkable?

However, in this coming biased competition, the human brain has the advantage of being swift and adaptable to the circumstances, but the computer is not left without arguments: it is never tired, it never dies, it does not belong to any syndicate, it confidently memorizes everything and, last but certainly not least, it has a command the human brain will never enjoy: the command "undo." This command makes it possible to go back to the past without any trouble. "Return to the future" is a real behavior for the computer — we could never expect that! What's done is done in a human life.

In the present debate about the future of the human brain, what could a scientific mind, but non-specialist in biology or artificial intelligence conclude? I am a simple physicist, curious and objective but hanging to the demonstration and keen not to drift into the utopian possibilities or science fiction. Many have given in to this temptation of drawing definitive conclusions from hypothetical facts which were considered to have been established.

This book is also aimed at being accessible to the ordinary reader without hiding behind obscure theories.

## About the Book

Both versions of this book will make a provisional comment and open a bit of thought about a future that is growing around us. Many of the topics can be addressed from very different perspectives, making a classification of them foolish. So you might find comments on the same subject in different places, from different angles. However, we will try to bring order and consistency to this flow of ideas, and this overview will be done in three steps.

First of all, we are going to explore the current knowledge of our brain; this organ remains by far the most mysterious in our body. Former President of the United States Barack Obama said on April 2, 2015: "[the] brain is the next frontier"; substantial efforts (and also heavy funding) are invested to understand how our "idea box" is working, how it could be possible to "put our fingers in", in order to fix, strengthen, improve, or entirely reshape its inner functioning. Of course, the more this knowledge sharpens, the stronger the impact on our fundamental being. This also is of manifestly indisputable strategic importance and the nation which manages to control any decisive process first will benefit, on every front (military, economic, social, etc.), a major step ahead its competitors; thus, we should not be surprised of the level of financing which the nations are devoting to research in the fields of biology and software science.

The second part will be devoted to the challenging development of the computer itself and its AI counterpart. Where are we now and what about the future? Computer technology is now well-integrated into our lives and its increasing pervasiveness continuously invades our daily lives, intruding into our privacy, accordingly changing our habits. The role of supercomputers becomes significant in every domain where the volume of data to be taken into account exceeds the possibilities of a human brain (health, transportation, banking, strategy, meteorology, etc.). At the same time, the software and corresponding dedicated instruments are a source of great hope for curing hitherto incurable illnesses (particularly brain failures) and consequently extending our life expectancy.

Last but not least, in the third part of this book we will examine how the two axes of research (biology and computers) could meet together and profitably cohabit. Is the computer to become a support, a complement, a reinforcement for the brain, or would the brain be doomed to behave as a slave of the machine? Would man become a cyborg? Could it be backed up in a copy on a computer (non-wearing, insensitive to neither diseases nor aging, duplicable … improvable)? Could mineral logic be compatible with the biological life or is a new logic to be invented which could involve biological circuits? Moore's law continues to raise the performances of microprocessors to unthinkable summits. Would a so boosted computer be able to become infused with the natural potentialities of the brain, such as intuition, abstraction, transcendence, and artistic or scientific creativity? Would there be a noticeable beginning in the intangible field of the arts?

Would we have to optimize the human brain to reach an ideal model (unique or multiple?) which would meet an ultimate eugenics? Would the biological man have a reason for existing when God would have disappeared?

It will be noted that, evidently, we will bring more questions that we will suggest answers, in this competition which has just seriously begun.

While waiting to compete with a human brain, the computer, in any case, is on the verge of greatly contributing to the exploration of the mind, in an effort where even achieving the first steps was unthinkable a few years ago. Now the leap forward in computing power allows scientists to tackle the corresponding huge amount of data, which led Joshua Sanes (BRAIN Initiative advisory committee) to say: "I believe, and many people believe, this is the biggest challenge of the century."

# Part 1

# About Our Brain
## How Does It Work?

Not so long ago, almost nothing was known about the brain except for its anatomy and some coarse electrical parameters. Barack Obama once said: "We are able to see the galaxies or very little things like the atoms but we still have a lot of mystery with these three pounds of nervous tissue between our ears."

Today, however, experimental means are there (and they get sharper every day) which allow deeper investigations into both the operation of the individual neurons with their global organization (in some way, the "hard"), and their programming (in some way, the "soft"), which has proven to be particularly complex and ever-changing. Does this organization, constantly evolving, result from randomness, some causality, or a superior will? Here we come to the realm of the divine.

A question also arises of knowing if it does make any sense to push ahead with such deep introspection. Already the Greek philosopher said "knows thyself", while exploring the philosophical logic, trying to understand its rationality. But how can an observer be able to consider himself as an object of observation, especially in the immaterial domain of thoughts? Is there some relevance in such an approach? Would the new science of metacognition meet the expectations?

Today, neurocognitivists (as neuro-psycho-electro-physio-cogniticologists are called in short) strive to identify and decipher the underlying mechanisms of our "self," in the undeclared aim to get a copy of it. Would it be a vain undertaking or would all of this finally reveal some terrifying truth? We honestly don't know, but we work with a tough obstinacy to understand what we really are. What is the precise meaning of the word "understand"?

To give the reader an idea of the level of complexity of such an undertaking, one can use an analogy: one can compare these researchers to a blindfolded man[1] who is to go shopping in a supermarket where he has never set foot!

The only resource he has is his white stick (which we will identify as "MRI", "EEG", etc.). This guinea pig will first have to find his way around and memorize the layout of the supermarket, which we may name "Consciousness." By trial and error, and also by intuition, he will attempt to analyze his environment and establish benchmarks. He will also meet (neural) employees who push carts around in order to supply the stalls; these employees exit from a huge warehouse (which we will name the "Unconscious"), entry into which is prohibited to customers and which is supplied from the outside by trucks loaded with goods (our five senses and the corresponding nervous system).

This work is obviously not easy for our blind researcher, who has to differentiate between a cornflakes box and a similar package containing laundry detergent, and to find the shelves with fresh eggs! And yet that is what he strives to do with patience and determination. What can we find on the shelves? How does the back room work, where the goods are identified, sorted, stored, and organized before being displayed? How are the pallets of merchandise removed from the lorries? How are they unwrapped? That is what we will try to discover, following, step by step, our blind man

---

[1] Throughout this book I will be using the shortcut "man" as a generic word for "human beings." Of course, women are included — who I dare not forget and whose brains work (to some extent) the same way!

in this evocative analogy, but knowing well that, in any case, his white stick will never allow him to read the price tag.[2]

The success of this scientific endeavor is still far from being secured, but there is no other way of proceeding.

Let us follow the white stick!

---

[2] However we can nevertheless imagine that the price tags are written with Braille characters and the white stick corresponds to fingers! In this analogy could the genes in the DNA molecules be comparable to Braille print?

# Chapter 1

# The Biological Hardware (Hardmeat)

R.U. Sirius[3] (Ken Goffman) wisely uses the word "hardmeat" to refer to the (somewhat soft) biological "hardware" which backs the brain's activity, and "softmeat" to refer to the way our brain is programmed to exchange and process information.

Of course, understanding the hard part of our brain is a tremendous task, but to make a copy also requires deciphering in depth how it works, i.e. the soft part, and that is not for tomorrow.

## What About the Brain?

This part of the book is not intended to be an academic course about the neurology of the brain but, more simply, to assemble some indications, some elements of knowledge which could assist the reader in following the twists and turns of the match between the human brain and the computer. However, one cannot avoid asking this central, inescapable question: in this competition, does the computer have to truly mimic the brain, or would it not be better to let it develop its own skills by itself, without the aim of copying the brain?

The brain, as we all know, is encased in a bone box we call the skull, which protects it from external shocks; but this precious and delicate

---

[3] *Transcendence*, R.U. Sirius and Jay Cornell, Disinformation Books, 2015.

organ has, in addition, to be kept far from direct contact with the bone wall. This is the role of the meninges — membranes which gently wrap the brain and act as a layer of protection together with the intermembrane liquids. With all this careful packaging, the brain is ready to do its fantastic and unfathomable job.

The size of the brain is in the range of a liter or so and it weighs, as stated by Barack Obama himself, some three pounds. Then is all said about this box which houses our self, our soul (?), our personality? Humph! Far from it!

Some people thought that our "intelligence" (to start with this word which we will meet throughout this book), or our "cleverness" if you prefer, is directly related to the size and shape of our skull. Roughly, the rule was: The larger the better![4] This assumption gave rise to a wealth of attempts at finding correlations also with the complexity of the structured external shape of the brain in the framework of a "science" known as phrenology. Of course, we get the information after the brain has been removed, and then not any conformity check of the intelligence can be undertaken afterward!

This approach was also tentatively proposed for detecting criminal birth trends. It was later generally agreed that all of that comes to nothing. The mystery of our brain's shape and size persists.

It must also be said that in that time the only available experimental means were reduced to visual examination. From the prehistoric ages men were curious about what goes on in this box and it is known that, in various regions of the world, people, even with the help of flint tools, opened the box in a tentative trepanation to see what was inside (not yet Intel, assuredly!). X-ray radiography was still far from being invented.

Now the situation is quite different and we can benefit from a wealth of investigative techniques which allow us to get precise and noninvasive information about the nature and the activity of the brain.

On top of that, it must be recalled that, as the computer does, the brain consumes a lot of energy; the brain is the more voracious organ concerning energy consumption, way more than the heart, stomach, or muscles.

---

[4]Of course, if this approach was stupid concerning human intelligence, it nevertheless obviously holds for AI and computers which advocate: "The bigger the better!"

Permanent consumption, throughout the days and the nights! This energy is supplied by the blood, which carries red cells full of hemoglobin and oxygen, which, in its turn, comes from the lungs. Should this permanent supply be stopped for any reason, the brain immediately disengages and you become unconscious before entering an irreversible death process if the oxygen access is not rapidly reactivated.

Surpringly, it is to be remarked that if a comparable activity was to be performed by an equivalent computer (current technology) the consumption of electrical energy would be infinitely larger.

## Gray Matter

The core activity of the brain — this is well known — is concentrated in what we call the "gray matter," so named because of its particular aspect. This gray matter makes up the superficial layer of the brain and covers the whole external surface; in order to increase this usable area, the brain is provided all over with deep, folded, and complex ridges. A theory was held for a long time that the more the lobes are pleated the more clever the person is, and that idiots have a flat brain! But it was afterward discovered that, as for the size of the skull, the relationship is not so straightforward!

This shallow layer, or "cortex," contains the body of the neurons which manage the information and convey it through their extensions (axons) to another part of the brain where they meet other neurons. The axons deeply penetrate the brain, sometimes through complicated paths and loops. Some zones are more or less specialized in a particular function, even if the total activity remains globally distributed.

The cerebral cortex structures are never identical in the detail but only look similar from one brain to another. The same holds for the brain as for the human face. Everybody has a nose, a chin, two ears, and so on, but these elements make a unique combination: our face. Same thing with our fingerprints, which drastically differ although they are very similar.

The general configuration remains the same from one brain to another; there are two main distinct lobes which are separated by the "corpus callosum," which places a communication interface between the two lobes and is traversed by the axons.

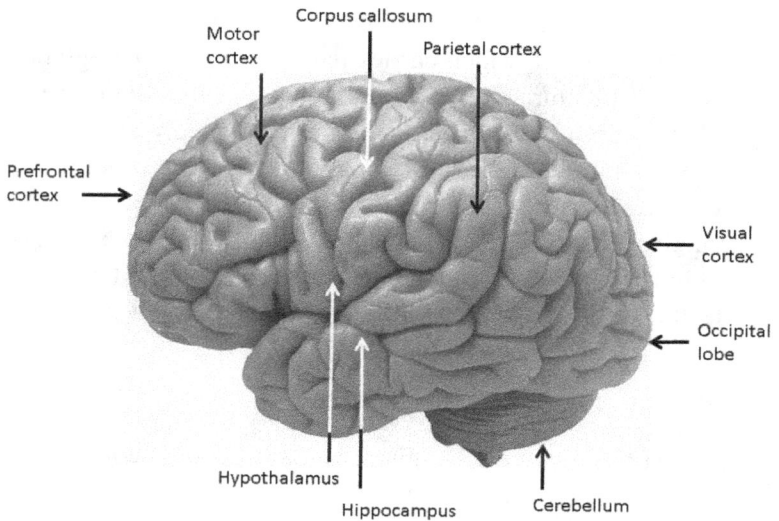

**Figure 1.1.**   The whole brain — an external view.

One also finds specific organs related to the brain (Figure 1.1), like the hypothalamus, the hippocampus, or the cerebellum, all necessary for the proper functioning of the entity. Each has its own role. The hippocampus, for instance, specializes in the establishment of long term memory after managing the short term memory, whereas the hypothalamus, associated with the olfactory bulb and the pituitary gland, is related to the endocrinal system.

Also, the brain cannot be dissociated in its operation from the spinal cord, either from the optical nerves and the retina or from the hearing nerve, which are all intimate extensions of the brain. All of that should be taken into account if one intends to make a realistic electronic copy of the brain.

## White Substance

Under the cortex, a white substance occupies almost half the volume of the human brain, much more than for other animals. This volume is larger

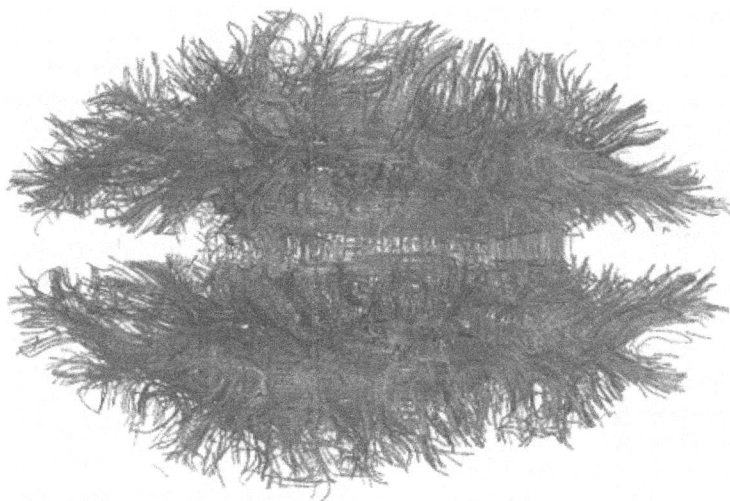

**Figure 1.2.** A view from the above of the paths of the axons through the two lobes of the brain as obtained by high resolution MRI. Of course, it is a purely graphical display and nobody yet knows or has afterward verified if it really represents the full reality.

for men than for women.[5] This white substance is made up of billions of axons connecting together the neurons in the bulk of the brain and within the two lobes (Figure 1.2).

These axons, connecting the neurons one by one, as electrical wires are insulated, more or less efficiently, with a myelin envelope; the electrical signals (ionic or electronic) propagate even better along these axons if the myelin layer is thick. This matter is produced by the "glial cells" and so provides an insulating layer for the axon, similar to that around an electrical cable. We still ignore what could be the mechanism which optimizes the thickness of such an insulating layer as a function of the length of the axon fiber in such a way that the signal can reach the very end without being too cumbersome in this limited zone of transfer.

---

[5] Which means that there is a natural and definitive difference between the genders, even at the brain level. Each benefits its own configuration and no one is superior or inferior, but simply different. Nature is the boss!

The axons are also packed in "columns," each with its specificity but respecting a common affinity. This organization has not yet been exactly deciphered. When such a set of axons is collectively stimulated electrically, the transfer of the electric charges generates a more intense electromagnetic signal, and this "coordinated firing" is intense enough to be recorded from the outside (such as a lightning strike).

This body of wiring is constantly evolving following us learning new things and corresponding reconfigurations of the neuronal links. In a healthy 20-year-old man, the cumulative length of these links reaches 176,000 km! That's a lot! With increasing age this length slowly shortens, particularly with the loss of the neurons whose axons are too short or weakly protected, but we currently know[6] that new ones are permanently generated from the hypothalamus at a reduced pace, and they are able to find their place in the brain according to their properties. They are especially required in the memory zones. The relationship between the detection of a local gap and the production of the corresponding specialized neuron is still to be discovered.

Physical disturbances which could occur in this volume of the white substance obviously dramatically impact the proper functioning of the "neural plant." We are still not able to master Alzheimer's disease or multiple sclerosis or other multifaceted degenerative neural diseases.

## Copying the Wiring

The complexity and upgradability of this bioelectronic structure is not so readily fully understood; thus, making a software copy of it is not so obvious, despite what Kurzweil said about it[7] who foresees such a completion within the next five or ten years at the most for a "transcendent man."

Such projects for identifying the wiring scheme (or "connectome") of the brain have been flourishing for years around the world, using different approaches. Some are trying to recognize and record the paths in a living

---

[6] "Marked loss of myelinated nerve fibers in the human brain with age", L. Marner, J.R. Nyengaard, Y. Tang, B. Pakkenberg, *J. Comp. Neurol.* 462(2), 144–152, 2003.
[7] *The Singularity is Near: When Humans Transcend Biology*, RayKurzweil, Penguin, 2006.

brain by means of high resolution MRIf[8] providing microscopic scale views; others are more simply interested in the functional and structural links, at the macroscopic scale, between the cortical areas and structures.

In the United States, the National Institutes of Health (NIH) coordinates and funds the Blueprint for Neuroscience Research project, the two poles of which are directed by Harvard/UCLA and Washington University/ Minnesota University with the contributions of the NSF, FDA, and DARPA.

A similar project, Whole Brain Emulation (WBE), is being pursued in Europe (Blue Brain Project), headed by Henry Makram. Located at the Polytechnic University of Lausanne (Switzerland), it is based on the Blue Gene IBM computer.

Another project, called BigBrain, is being carried out in the same organization as Blueprint. It aims to extract information on the brain through frozen and "microtomized" brain slices and 3D microscope examinations! This is quite foolish; we will address this concern later. A simpler model might have been obtained for the worm *C elegans*, which has only 302 neurons in its brain.

Inevitably, all these projects encounter the challenge posed by the considerable volume of the connections to be followed ($10^{14}$ links in a permanent reconfiguration) and the impossibility of individualizing each synapse of each axon of each neuron. Let us recall, as a comparison in the domain of large numbers, that the number of base pairs in a human genome, so difficult to be numbered, is "only" $3 \times 10^9$!

All this effort of reverse engineering requires tremendous funds from different sources. Not to mention the similar efforts carried out in other countries, such as China.

Anyway, it must not be forgotten that, even if a complete and reliable wiring scheme of the brain connections (possibly frozen) is obtained, that should not mean that is how the brain works and how it evolves, because the brain, like every other organ but probably more so, reconfigures itself constantly.

---

[8] Magnetic Resonance Imaging (functional). High resolution is obtained by increasing the magnetic field (currently 4–11 teslas in the latest equipment).

Our brain today is different from our brain yesterday, with regard to the pace of our instant life. If a computer copy were available, it should necessarily benefit from certain autonomy of evolution unless it was permanently linked to the original in a slave process and follows all of the evolutions. Then, if the copy were autonomous, it would certainly like to "live its life" and soon diverge from the original.

Supposing we are able to copy a brain, which one should we choose? A normal one, we may say. But what is "normal" in a brain? Would it be normal to choose a normal brain or would it be better to choose an exceptional brain? And what to do with the copy? A simple backup just in case? Or a basis for improvements? The questions inevitably arise without any straightforward answer.

## The Basic Cell: The Neuron

We have already emphasized the way the neuron operates; let us now enter into the details of this particular cell.

### *The Constitution*

The neuron is made up of a central cell with its nucleus containing the DNA molecule, and:

- Long tails or axons (some microns in diameter), often unique, which convey the action potential toward the outside; sometimes very far through the brain, following complex paths; the neuron is the longest cell of the human body (up to 1 m);
- At the end of the axon we find bouquets of synapses which form some 10,000 nanometrical sensors and are in ionic contact with the next neurons;
- The dendrites around the nucleus (some 7,000) receive the input signal coming from the other neurons, as sketched in Figure 1.3.

Each neuron differs from the others (even genetically). They differ in size or general constitution but also have different functions.

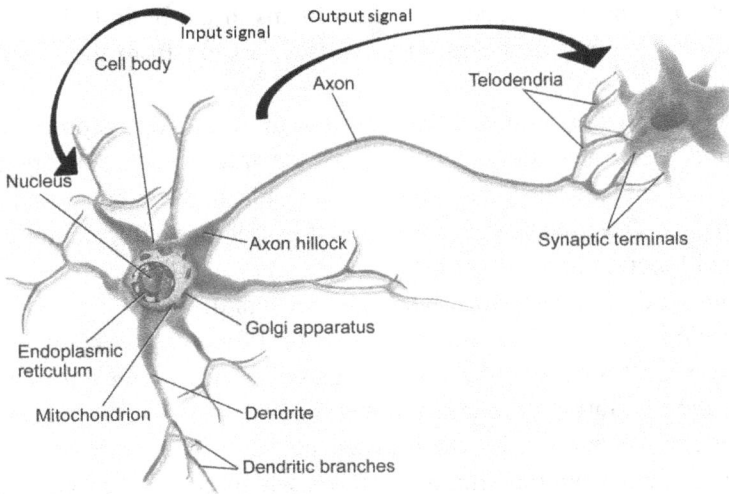

**Figure 1.3.**   Neuron structure — schematic.

For instance, the neurons which belong to the prefrontal cortex (a superficial region of the front of the brain) contain a specific gene (named FoxP2) which generates advanced dendrites and are especially localized in the speech zone. Others are devoted to particular recognition processes. This is an uncrackable proliferation.

The neurons obviously can be found dominantly in the brain itself (some 100 billions), but also in the spinal marrow or even the intestines (100,000).

Giant neurons are also able to convey conscious information throughout the brain; their axons cross the white substance from one lobe to the other. Such architectures of specialized neurons have already been modeled in computers.

The Blue Brain project (Henry Makram) already proposes reverse engineering of the mammalian brains, whose basic functional units are the neocortical "columns" containing the longest neurons.

Neurons which are not able to provide enough connections (say, 1,000) are immediately induced to commit suicide and are eaten by neighbors. When such disorders become too frequent, they are not spontaneously cured (aging) and may induce severe problems in the brain (Alzheimer's, Parkinson's, or other degenerative processes) arising from the functional disruption.

It has long been supposed that neurons are not able to reproduce and so are not regenerated. One currently knows that (contrary to the monkey) this loss is compensated for; neurons are able to be regenerated, at a reduced pace, from the hippocampus and the olfactory bulb, and take their place and function in the right area.

Neurons do not reproduce but brain cancers do exist, which implies uncontrolled reproduction. However, this proliferation does not involve the neuron itself but rather the neighboring cells, such as those from blood vessels, the pituitary gland, external membranes, etc.

Neurons can modify by themselves their own genetic heritage. The neuron contains scattered strands of DNA which can reaggregate to the main molecule in a random operation; then they migrate along the chromosome to keep a stable place (they are called "jumping genes"). Then the brain can be considered as a random GMO plant! These new neurons consequently find their corresponding place and function. The neurons do not reproduce; they die hard but newcomers are continuously regenerated: the neurons are in a perpetual activity of rearrangement.

The neurons of the same person may have different DNA and evolve in a more or less causal behavior. Intelligence depends not only on the genes that were passed down by the parents and the cultural context but also on random and poorly understood elements (random and necessity) — some say destiny! The diet, the way of living, and the environment play a role as determinant as the genes in the brain's organization. These factors are sometimes strong enough to induce specific modifications down to the genetic expression.

The manipulation of the genes inside the neuron starts to become possible through optogenetics and atomic force microscopy (AFM).

Neurons are specific to particular functions, which means that the human cortex is still more complex than was initially foreseen, with hundreds of different types of neurons in a population of 100 billion neurons in a single brain!

Cloning, grafting, and reconstruction of a brain is assuredly not to be expected tomorrow. Be that as it may, therapeutic cloning would assuredly be the best way to reinforce or repair a wounded brain (following the overuse of Ectasy, among other possibilities).

- French–Belgian researchers nevertheless succeeded in fixing the brain of a mouse (visual cortex), using embryonic stem cells which were mutated into neuronal cells, while a United States–Japanese team of researchers succeeded in cloning cells to cure Parkinson's disease.
- A partial graft of human neurons was also attempted at a hospital in Rouen, France.

Nevertheless, despite the fast and extensive progress of biology, growing an artificial brain will not be so easy.

### Biochemical Activation of the Neuron

Every biological mechanism which manages the neurons obeys strictly the laws of physics and chemistry, but we still must know which ones.

Some inexperienced philosophers ultimately refer to quantum mechanics, Schrödinger's equation, Heisenberg's uncertainty principle, and so on. I guess they never viewed up close any Hamiltonian operator!

For a long time, it was believed that doping and enhancing the intellect might be possible using drugs or nutrients. Actually, there do exist stimulating elements whose effects are well known, but they never actually generate geniuses. On the contrary, some other drugs (Ectasy, among others) are well known to definitively destroy the neurons.

In addition, trials were conducted using transcranial magnetic stimulation (TMS), which provides the brain with magnetic field impulses at a very low frequency, focused willy-nilly on the frontal cortex.[9] Unfortunately, the results were never very convincing.

It was also observed that the neurons could be driven by the light of selected wavelengths through very thin optical fibers using the photoelectric effect on individual molecules in the near field range.[10] So doing it could be envisioned to deconstruct the neuronal circuit of Parkinson's

---

[9]This is typically a "white stick" process!

[10]*Near Field Optics and Nanoscopy*, Jean-Pierre Fillard, World Scientific, 1996.

disease as well as with electrical probes. However, there is still the una-voidable issue of the dimensional hindrance; even if the fibers are actually very thin, they have nevertheless to find a way in the gray matter, and it never happens without creating local damage, especially if the operation has to be repeated several times at several places. Also, how to guide the fiber at the nanoscale remains far from obvious.

The ionic equilibrium of the neuron provides a resting static potential in the range –60 to –90 mV. The nervous influx reaching +35 mV allows a modification of the permeability of the cell membrane, which induces a migration of the electrical charges along the axon.

## Redundancy, Troubles, and Repairs

The huge number and variety of neurons offers the possibility of organizing a large redundancy in the internal communication within the brain. Such a mechanism is absolutely required to provide appropriate safety and to guarantee consistency of the brain.

If a problem were to occur, such as a cardio vascular accident (CVA), which is a rather frequent situation, or an accidental wound which harms the cortical tissue, the brain can find (to some extent) a way to recover by itself, by either generating new neurons or bypassing the damaged region. Efforts are ongoing to understand and stimulate this natural process with the help of adapted drugs such as anticoagulants. Of course, transplants, grafts, and so on are also emphasized.

At the moment, even if it is of the utmost urgency, a complete graft of a brain is not yet ready to be attempted! (Even though some monkeys have already been sacrificed).

Among the various kinds of neural diseases, Alzheimer's is the most frequent and deadly. About three years ago, I asked a famous geriatrist: "Is there hope, with the new technologies, to discover how to cure this plague?" His answer was unambiguous: "We don't know what it is, or where it comes from, we have not any precise diagnostic, we don't know how it evolves, there is not any drug available, so no solution can be expected, even in a remote future." A year ago, I was told that a protein which has been discovered could be at the origin of the disease, and this week I read in *Fortune* that we are now able to precisely diagnose this specific scourge and that experimental treatments can now be proposed (in limited cases, but it is just the very beginning).

The current explanation involves "$\beta$-amyloid" protein molecules which accumulate in plaques along the axons and hamper the electrical transfer. Two ways are being explored to tackle the problem. The first one, of course, is to chemically destroy these deposits; the other is to enhance the natural reaction of the neuron so as to prevent this layer deposition. That work is underway and no doubt improvements will soon be found.

Another unexpected phenomenon has been discovered very recently which is worth mentioning. Hearing impairments often occur (with aging) which are not easily identified at first and may induce deeper problems if not cured. MRI revealed that the corresponding hearing areas in the brain progressively shrink due to the lower activation of corresponding neurons. This is, in some way, instinctively compensated for by enhanced recourse to the site where the brain area is correspondingly stimulated to grow. As an indirect consequence of this plastic restructuring, the Alzheimer's process could be reinforced. With the help of hearing prostheses it was established that the process could be slowly attenuated or even reversed.

These examples clearly show the fantastic speed at which things are changing in the exploration of the brain.

## Biological, Biochemical, and Electrical Logic

Electrical or chemical transmission of the information from one neuron to another (or several) is managed through the synapses in a very diversified manner, by using specific molecules called neurotransmitters. A neuron fires when the membrane potential reaches a certain level. It is likely that the operating model must include delays, and nonlinear functions, and so obeys a set of differential equations describing the relation between electrophysical parameters such as electrical currents, voltages, membrane states (ion channel states), and neuromodulators.

The neuron is in no way only a passive relayer of transmission, but it is also an active center of information processing which takes into account the multiple messages it simultaneously receives, and immediately gives a convenient answer.

Then, from all of that, enough is known to say that it does not amount to very much! Actually, we know how the neuron works electrically when it is fired to pass a signal — but to which other neuron and why this one? The underlying determinism, however, remains elusive. From where does the decision of the stimulation come? A coincidence of several signals? If the role is purely passive and determinist as an IC would do, what is at the origin of the command? Are there active neurons which create information? Would there occur a random trial/error and probabilistic mechanism?

There are the questions an enquiring and nonspecialist mind could raise. That's right, referring to the word "unconscious" to explain that permanent activity around the clock. Who cares? How would it be possible to enter all that stuff into a computer to mirror this activity?

On top of that, it can be added that the nervous system in the body is provided with a warning system to detect any foreign or perturbing element which could cause interference or hazards: this is pain. Corresponding signals are quickly identified and localized by the brain, which triggers a timely and well-targeted response. One could dream, of course, of a pain-free life if these local or cerebral nervous centers were disconnected, but this certainly would be very risky because we would no longer be alerted of any dangers and so prevented from reacting in due time.

In a simulation of a brain, would it be necessary to also make the computer sensitive to pain? In any case, the computer does not care about the very notion of a potential danger or felt concern; things might be changed if a model of the notion of consciousness could be simulated in a computer.

All these questions are preliminary to the establishment of a "thinking machine" representative of the reality. The formal neuron is expected to reproduce as closely as possible the biological one, that is supposed to be standard (it being understood that a diversity of models will be the second approach, if it is intended to come closer to an artificial brain).

This did not prevent mathematicians, for a long time,[11] from elaborating models from diversified logical functions presenting multiple accesses

---

[11] "A logical calculus of the ideas immanent in nervous activity", Warren S. McCulloch and Walter Pitts, *Bull. Math. Biophys.* 5, 115–133, 1943.

and unique output and assembling them in learning networks for an auto-matic adjustment of the coefficients. It is in no way claimed that these various models of artificial formal neurons would truly mimic the bio-logical neurons; the intention is only to evaluate the properties and behav-ior of some mathematical functions in order to obtain efficient networks.

## Transplants

In order to be exhaustive with the constitution of the brain, a terrific opportunity has just appeared which takes us into the domain of Frankenstein: I noted from a press release[12] that an Italian neurosurgeon (Dr. Sergio Canavero), in a research group from Torino, proposes to make a complete head transplant and is looking for "donors" (and also for fund-ing from American sponsors). In this offer it is not clear if he is looking for a head to graft onto a body he already has or the opposite. In any case, what to do with the remaining corresponding body and head? Would they also have to be grafted together? Would he be grafting a head onto a body or the opposite? Supposing that this works, what about the benefit of such an exchange? That is definitely a crazy plan, but what is fantastic about this story is that people are seriously busy with it!

I wonder (I am not alone) how it could be possible, even with every conceivable technology, to connect all nervous links, to reconstruct the spinal cord and the backbone, and the complete throat, all of this without interrupting the blood circulation. In case the operation is successful, who would finally be who?

Of course, such an imaginary possibility in grafting has been known in science fiction literature for a long time, to the point where Bill Gates offered his head for such a project (when he dies, I guess).

# Methods of Analysis

For a long time, people were looking for physical means to explore the brain — unsuccessfully until recent times, when experimental tools began

---

[12] http://www.midilibre.fr/2015/06/13/le-neurochirurgien-qui-veut-transplanter-une-tete-humaine-en-appelle-a-bill-gates,1174702.php

to be available. In spite of that, the brain remains a deep mystery to unravel. The human cortex is still more complex than was initially foreseen.

Brain researchers generate 60,000 papers per year. "But all focused on their one little corner. ... The Human Brain Project (HBP) would integrate these discoveries and create models," said Henry Makram, who has earned a formidable reputation as an experimenter. So what are the means already available? How to access the knowledge of the brain?

## *Direct Investigations*

Currently, we do have very sophisticated and effective means of beginning to get information about the way the brain operates, even with the help of detailed three-dimensional digital images (thanks to the computer!).

Some 50 or 60 years ago, a simple X-ray radioscopy constituted quite an accomplishment which required putting the patient to sleep. The main difficulty was that the brain is surrounded by the skull, which is made up of bones. These bones strongly absorb the X-rays, whereas the soft tissues in the brain do not. The contrast required for a clear image is then very difficult to achieve and the resulting photograph is poor even if the X-ray dose is high. This has, for a long time, been a problem in getting a meaningful image of what is going on inside. The only way forward was to chirurgically open (trepanation) and look at the external aspect of the brain. Of course, there were no means to go inside without dramatic and definitive damage. Currently, things have moved and largely improved; we will come back to the radiologic means a bit later.

### *EEG and related devices*

Another experimental approach, still widely used nowadays in a better-performing manner (thanks to dedicated computers), is electroencephalography (EEG). It was discovered a long time ago (Richard Caton, 1875) that the internal activity of the brain generates an electromagnetic activity which can be recorded externally; however, we had to wait until 1950 for the EEG technique to be actually used.

There are currently five identified and different types of cerebral waves: delta, theta, alpha, beta, and gamma. The main advantage of EEG is that it is very cheap and easy to use and is absolutely noninvasive. Of course, the waves have to be amplified and carefully filtered to be selected; this is particularly easy with the current digital electronics.

The time resolution is satisfactory. However, the space resolution is poor, and so it is rather difficult to identify the emitting region with some precision. Then the individual neurocognitive processes can hardly be isolated.

Nevertheless, it must be added that electronics and computer-assisted intelligent analysis of the signals continually improve, and allow a better understanding by correlating the signals recorded simultaneously from many probes stuck at different places on the skull. Up to 256 electrodes are usually put into action to get more detailed information. EEG is often used in conjunction with other means of analysis (CAT, PET scan, MRI, etc.) to reveal correlations and remove ambiguities.

External stimulations of the patient also help trigger the corresponding waves (flashlight stimulation, twinkling of an eye, speech, etc.). This kind of analysis is of primary importance in detecting consciousness phenomena related to subliminal processes; a back-door to the unconscious domain in some way. Recently, Oxford University researchers succeeded in getting personal information from the brain of a volunteer by combining EEG with a dedicated question procedure.

Also, some classical brain diseases, such as epilepsy, leave a typical signature on EEG records, which allow a clear identification and diagnosis. Cerebral death, coma, confusion, and cerebral wounds are also readily identified through EEG.

There are various types of cerebral low frequency waves permanently active in different parts of the brain:

- The alpha rhythm spread over the 8–12 Hz range of frequencies with displayed voltages in the range of 25–100 $\mu$V, in the occipital regions;
- The beta waves are faster (13–30 Hz) but of weaker amplitude (5–15 $\mu$V) in the frontal regions;
- The theta waves can be found in the frequency range of 3–8 Hz;
- The gamma waves are in the high (?!) frequency domain above 30 Hz.

But special bursts of activity also occur when the brain is prompted by external stimuli (speech, image, and so on), especially the P300 wave, which is associated with the "global ignition" of consciousness (Figure 1.4).

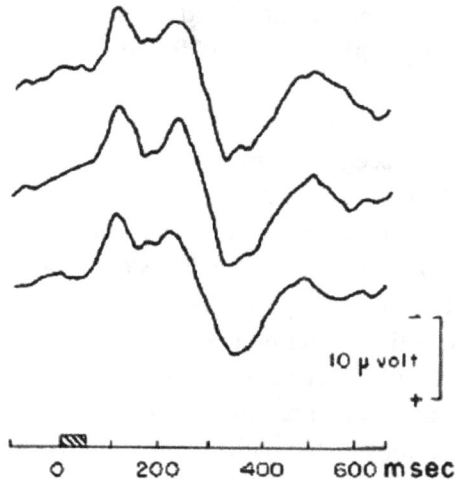

**Figure 1.4.** The P300 burst wave.

The skull itself is not the only place where neural electromagnetic (e.m.) signals originating in the brain can be detected; electrodes can be placed along the paths of the nerves toward the muscles, for instance along the arms.

In a reciprocal way, such electrodes can be used to stimulate the muscles with an artificial signal generated after a brain command is detected on the skull, in order to bypass a wounded spinal cord. This method of "thought command" was recently used to help an individual with tetraplegia to use artificial hands. The brain signals were simply provided by external modules placed on the skull at the right places. This method avoids the requirement of implanting an electronic chip inside the brain, as was done before. We could consider, in a not-so-remote future, piloting in this way a complete exoskeleton or a full humanoid robot (such as a military one?).

Of course, the use of such a thought command is also emphasized in many other applications involving a robot which can directly be activated in the blink of an eye!

This technique of recording and analyzing the electrical activity of the brain actually benefits the development of miniaturized electronic chips.

The latest gimmick is an "electronic skin" (Biostamp[13]) the size of a small postage stamp that can be stuck on the pinna of an ear or elsewhere; it could be able to track brainwaves in real time and transmit them as identified messages through a computer spelling interface, or to detect seizures in epileptic patients, or … to pilot drones! Optionally, the chip could be implanted and it would automatically dissolve after a programmed delay. One problem still remains unsatisfactorily solved, as in any wireless system: the battery!

In some cases (such as Parkinson's disease or other neural conditions), it becomes necessary to introduce electrodes or optical fibers directly into the brain to reach a sensitive point where an electrical or optical stimulation is to be locally fed. This is not so easy to do but it works.

Of course, the e.m. waves emitted by the brain are only a consequence of the collective neural electrical activity; they do not give any information on why they are emitted: Then they would not in any way be a reliable method of communication.

However, in spite of that, the shape of the signal also with the brain region involved makes it possible, in some cases, to reproduce the very meaning of the brain message. This is called "subvocal speech." The Audeo system, for instance, uses a lightweight receiver placed on the neck of the subject, which intercepts the signals the brain sends to the throat to generate speech. Then the signals are encrypted and transmitted wirelessly to a computer which is able to identify the words through the individual phonemes and reconstruct a phrase in a similar way to a voice recognition software program. The message can then be processed to command a robot (a wheelchair, for instance, or a cellphone, or a simple set of speakers). NASA is developing similar systems of voiceless speech for astronauts. In some way this is the voice of the thoughts! Only the voice, not the meaning of the thought itself!

In 2015 Mark Zuckerberg (Facebook) said that "[t]his will be the ultimate mode of communication" which could read minds. This, of course, opens the way to significant drawbacks as long as "brainjackers" can take advantage of such an opportunity.

---

[13] "Electronic skin makes your body a computer", Kieron Monks, *CNN*, May 15, 2015. Available at: https://edition.cnn.com/2015/05/15/tech/electronic-skin/index.html.

## X-ray-related methods

Since the beginning, X-ray analysis has been greatly improved with contrast-enhanced and X-ray converter photographic plates, but the real progress came from computer-assisted tomography (CAT). A battery of focused X-ray beams and corresponding detectors are swiftly rotated around the head (or the whole body) of the patient and the data recorded, just as swiftly, in a computer which reconstructs a virtual image of the irradiated slice of the brain, and then slice after slice a complete 3D scan is achieved. The computer is able, with this volume of data, to determine the X-ray absorption coefficient of each elementary volume in the brain (or voxel) and thus recreate a 2D image of tomodensitometry of the inner part of the brain, in any direction, which makes a complete 3D exploration feasible.

The progress of this technology in the last 30 years has been astonishing, in terms of speed of scan, geometrical resolution, dynamic range, precision, X-ray exposure limitation, and graphical assistance. CAT images are really impressive, considering the difficulty of the operation.

Of course, the only valuable information brought by these images is purely morphological; biological information is more limited.

## Magnetic Resonance Imaging (MRI)

The real breakthrough in brain examination came with MRI, which is the fantastic technological implementation of a nuclear-physical phenomenon — the magnetic resonance induced in the magnetic moment (spin) of the protons of light atoms such as hydrogen or oxygen atoms, which are rather abundant in the biological tissues.

The patient is introduced into a tunnel where a powerful magnetic field is created by superconducting coils in order to orient the atoms in the same direction in spite of their natural thermal agitation. This field is complemented with another steady field, which can be oriented in the three directions ($X$, $Y$, $Z$) in order to create an orientable gradient to control the "slice" thickness.

A radio frequency wave is then applied to make the magnetic moment rotate around the field axis in a precession movement. The choice of the frequency is typical of the atom which is to be revealed. When the radio frequency is stopped, the relaxation of the atoms gives a back signal whose relaxation time depends on the biological environment of the atoms; then information on the density of the corresponding atoms can be deduced, whereas the position in the field gradient gives the localization. A real 3D tomography is then obtained by the computer, which is specific to each kind of atom; this makes it possible to identify biological species of molecules such as hemoglobin, which is a main tracer of the brain's local activity.

This procedure may seem pretty complicated but assuredly it works quite well and the images are beautiful and detailed, with their coded colors corresponding to the various specific tissues detected. Also, MRI is noninvasive and, contrary to CAT, perfectly irradiation-free.

Of course, the stronger the magnetic field, the sharper the definition of the images. Mini or even nano-MRI can be performed to get internal images of some animals or even individual biological cells such as spermatozoids, using a smaller experimental setup which can provide us with higher magnetic fields (up to 17 teslas) and thus reach the domain of higher resolution.

The improvements obtained for such MRI apparatuses currently allow us to get volumic images in quasi-real time, thus making it possible to follow in detail the blood circulation in the brain. This is the basis of what is called functional MRI (MRIf), which makes possible a real video of brain life. MRI is by far the most powerful and useful tool for exploring the brain, even though it is rather complex to use and not cheap.

## Positron Emission Tomography (PET)

This is also a method originating from nuclear science which allows, by scintigraphy, detecting in the volume the metabolic or molecular activity of an organ; it relies on the emission of positrons (antielectrons), the annihilation of which gives rise to two photons. The coincidence of these emissions allows pinpointing of the exact place of the positron emission and the density of the radioactive tracer (carbon, fluorine, nitrogen, etc.) previously injected into the organ.

Then PET is able to image the metabolism activity of the cells; it is a functional imaging technique which enables a possible diagnosis of pathologies like cancer. In 2014, a specific procedure was proposed in neurology and psychiatry, for example to detect the amyloid plaques on the axons of the neurons which develop in some cases of neural diseases such as Alzheimer's.

This is a very promising technique for studying the mechanisms of the neuro-receptors involved in the cerebral activity. The diversity of the possible tracers to be used can help make a very specific molecular detection in the cells. New equipment is on the verge of being proposed which combine PET with MRI. No limit!

### Magnetoencephalography (MEG)

This new technique of brain investigation and imaging is still experimental; it is not yet in routine use. Like EEG, it is perfectly noninvasive and relies on the emission of a magnetic field associated with the local electrical currents which come with the brain activity. Of course, the magnetic signal is extremely weak and cannot easily be separated from the ambient noise. Only the strongest currents can be detected; they arise from the collective and synchronous action of several thousand neurons, such as those involved in the components of the P300 wave associated with consciousness.

This magnetic signal, unlike the simultaneous electric signal of EEG, crosses the skull without any distortion, which makes easier the localization of the emitters. However, it was not until recent years that a convenient detector of the magnetic field became available: the superconducting quantum interference device (SQUID), which, as the name suggests, requires superconducting material and temperatures close to absolute zero (like MRI). The system also has to be installed in a specially designed room whose walls shield one from external magnetic fields (the geomagnetic one, *inter alia*).

With all of these careful precautions and a complex computer treatment of the data, high resolution images are obtained which significantly complement the paraphernalia of the other investigative means.

## Stimulation Methods

Another way to learn about the brain is to introduce a perturbation from the outside and look at what happens (always the "white stick" method).

### Electroconvulsive Therapy (ECT)

The most ancient and famous method is electroshock or electroconvulsive therapy, which dates back to the time of the discovery of electricity. The electrical discharge triggers a convulsive seizure which could result in curing certain psychiatric pathologies. The details of the operating mode in the brain are still unknown and currently this is a "last chance" therapy. This method does not provide additional information on the brain's electrical constitution.

### Transcranial Magnetic Stimulation (TMS)

This uses an impulse of a focused magnetic field to induce a local electrical perturbation in the cortex. The field is able to locally depolarize the neurons, thus modifying the local potential, which migrates along the axons toward the central motor pathways. This is a way to investigate the brain injuries following an AVC or multiple sclerosis.

### Pacemakers

Brain pacemakers are used in Parkinson's therapy, which proceeds from an electrical probe introduced into the brain to generate a deep or cortical electrical inhibition of the active neurons into the identified zones.

## Disturbances in the Normality

To bounce back from disturbances observed in the "normality," chirurgical operation can also be performed as long as the origin of the damage is convincingly identified and localized.

Sometimes, dramatic errors have occurred, such as in the controversial lobotomy formerly performed to cure sexual disorders, often with catastrophic consequences. A famous case was that of Rosemary Kennedy (a sister of the U.S. President John F. Kennedy), who underwent a lobotomy at the age of 23 because of her disordered sexuality. The operation left her permanently incapacitated.

## Internal Communication in the Brain

The internal communication system in the brain connects the different areas of the cortex where specialized operations are conducted. Each of these zones has a more or less dedicated role and corresponding specific neurons: memory, attention, perception, awareness, language, and consciousness; not to mention visual, motor, olfactory, and other cortices distributed in the frontal, temporal, parietal, and occipital lobes.

The nervous extensions allow connecting these areas of the brain with peripheral nerve endings, and for a long time people sought such correspondences in order to have a means to "put the fingers in the brain" — that is to say, to cure certain troubles by reverse actions inside the brain. This was very empirical but sometimes there was some success, and such "alternative therapies" are still proposed. The more ancient one is acupuncture, which more recently was competing with auriculotherapy restricted to some specific points of the pinna.

As a conclusion of this brief survey of the "hardmeat," there arises a series of questions which are still pending.

What is to be taken as good, useful, essential, or only important in these tangled circuits? How do they work in detail and why? That remains to be discovered. But extending the knowledge by extending the experimental means and corresponding explorations — this is the road that we should concentrate upon.

# Chapter 2

# The Biological Software (Softmeat)

Now that we know everything about the brain's constitution, we should try to understand how it works. Obviously, the underlying "software" is as diverse and changeable as the "hardware" was. A lot of people since ancient times were expecting to put some rationality into this research in a highly reverse processing manner and with the only instrument being their own brain. One point, however, is to be taken as certain: the brain is an electrical device. Electrical charges (electrons as well as ions) migrate, condense, and so build up local potentials (some say "evoked" potentials). Basically, there is nothing to prevent a direct comparison with a computer, all things being equal.

Yet to be found: the software inside the machine.

## From Thales to Freud

For centuries, man has unsuccessfully attempted to decipher the biological software program which runs his behaviors and destinies. Unfortunately, in this particular issue soft and hard are intimately mixed, which creates quite a difference from computers. The real support of the "soft" in the brain (some say "softmeat") is still unknown; there is no hard disk and no software program to explore. Another point remains to be clarified: would the research have to be devoted to our own brain or the brain of some other human specially selected for the purpose?

29

The huge diversity of brains makes each brain unique and permanently changing — where and when to take the snapshot? How to make a choice? Some are reportedly simple minds; others are out of the normality (as far as this word can be precisely defined). Who could be worth being copied? Would we prefer to make a mix?

The paths a brain uses to think are unknown and the possible conclusions are surrounded by uncertainty; deciphering a brain faces a moving context. Rigorous scientific analysis is of no use if measurements can directly yield a less questionable truth.

The wiring of the brain, as said before, is inextricably complex and the acting software is distributed in all places. Also, a large part of the "soft" is hidden in the unconscious domain of the brain, where access is hampered. Redundancy and parallel operations are everywhere. It is, then, not easy to draw conclusions and know where the commands are located and how they could be activated in detail, from an electronic point of view.

Nevertheless, the first attempts of investigations were empirical and, surprisingly, organized by philosophers whose only means were their own minds. In those times there was not any computer to compete with humans! Philosophy, to some extent, could be considered as a reflection on the intellectual mechanisms and, in the light of this, as revealing underlying software.

## The Greek Philosophy

This informal approach can hardly be considered a science in itself, because it is in no way concerned with measurements. Philosophy is not an experimental science (even though attempts were later made, unsuccessfully, by Leibniz to create a logical and universal means of calculation), in the sense that philosophers do not make experiments themselves but only comment or meditate on the experiences or concepts of others.

Philosophy covers a large and diversified domain (I would say everything conceptual), some of the topics of which are related to an analysis involving the brain mechanisms even if there is no direct access to them. This spreads from epistemology or anthropology to language by means of studies such as logic, metaphysics, or knowledge theory.

The initial, consistent efforts to try and figure out how our mind operates unquestionably belong to the ancient pre-Socratic philosophers, such as Thales of Miletus in the seventh century. This legendary figure, who did not write anything, is renowned for (besides his famous mathematical concepts) his proposed method for analyzing reality, which anticipated scientific thinking. This was the first known attempt to clarify our thoughts and organize them in a rational and logical way, deliberately ruling out the mythological. Thales was later considered the precursor of the "seven seers" (not to be confused with the seven dwarfs!). He cultivated the spirit of the eternal soul.

Thales was the author of a profession of faith shared by the philosophers and the "scientifics" with the same viewpoint: theoretical (ἐπιστήμη: *épistémè*), practical (τέχνη: *tekhnè*), or natural (φύσις: *physis*), "Know yourself and you will know the Universe and the Gods." He was the first to propose that nature does not derive from an exogenous divine source but is inherent in the life mechanism — that is to say (this is not said yet), biology. But it was from the pre-Socratics like Pythagoras that the words "philosopher" and "philosophy" arose (φιλοσοφι'α: philosophy; love of wisdom).

This trend toward rationality and search for "mechanical" explanations went against the Greek cults, which were seen as a major social duty resulting from the legends devised by Homer, the spiritual father of ancient Greece. These cults did not fit into the framework of a genuine religion but were part of a mythology of anthropomorphic gods, despite the harsh criticisms of Xenophanes.

In the fourth century BC, there came philosophers called "peripatetic," because they developed their philosophical dialogues when walking: essentially Socrates, Aristotle, and above all Plato, who left us imperishable written materials. At that time the point was to agree on the description of the way the mind proceeds, the way the ideas arise and are managed in the various situations where individuals or societies are involved.

One looks for compatibilities, convergences, and similarities to reach a "philosophy" — that is to say, a conviction that is soon called "knowledge" (λδγος: *logos*) with this fundamental question: what could *logos* be? Surprisingly, this kind of philosophical debate performed by the peripatetic can still be found today, *mutatis mutandis*, in advanced discussion between specialists. Of course, the current topics were out of reach for the Greeks but the method remains the same. In similar fashion Brockman[1] proposes a debating contest between renowned scientists on the subject "would our mind be digital?" *Nihil novi sub sole.*

Of course, these ancient philosophers did not have any physical instruments to analyze the brain reactions, as is the case now. Their only art, and they were excellent in this domain, was mechanics as a completion of logic and reasoning. Pulleys, siphons, yokes, and keystones would be models for the renewal of the "forces" in the 16th century.

The whole philosophy was a matter of discussions which strove to put rational ideas in order. Logic became a new and essential field. These processes remain purely literal and do not consistently get closer to the medicine (it will take much longer), but that did not preclude reaching general conclusions which still hold.

"See and know, know and touch, hear and know" led to "Know and be acquainted with." Aristotle believed that we benefit from an intellectual intuition (*nous*) which was adapted to the acquisition of unprovable principles but, nevertheless, this inductive approach does not exclude logic. The inference theory is able to distinguish the correct types (worth being remembered) from the incorrect ones (to be rejected) in the argumentation. The discussion (that is to say, the ideas intermingling) contributes to this analytical approach to the brain's operation.

In the debate, the sophists upset the game with critical peripheral discussions which made the problem still more opaque. Their contributions — sometimes positive, sometimes misleading — guided the reasoning toward famous paradoxes such as "Let us suppose that everything I say is wrong; what would be the value of the proposition: what I say is wrong?" How to explain this mess to a computer?

Logic is essential to the brain's function; one is relying on a system considered reliable to try grappling with what is acknowledged and recognized; that is to say, building the reason (λδγος: *logos*); that is to say, knowledge as originally Plato intended. But the rhetoric questions the

---

[1] *Life: The leading edge of evolutionary biology, genetics, anthropology and environmental science*, John Brockman, ed., Harper Perennial, 2016.

very meaning of *logos*. Then, is everything misleading in our judgment? It can also be added that logic encounters fundamental difficulties from disturbing features of the language; the concept of truth, which is the basis of any scale of values in the logic, is seriously undermined.

Two of the essential axes of the reasoning, namely logic and the perception of the reality, prove deficient because of their variability and the lack of certainty and reliability.

All of that procedure aimed (already) at improving man by bringing him "wisdom" through a better knowledge of oneself and of nature. The perfection of the reasoning would have to result in an explanation of the world even if some were still intuitively defending the possibility of virtual worlds beyond our understanding. The "non-existence" of things was always the focus of bitter discussions. Parmenides defended the stand that it was unthinkable, whereas Gorgias[2] said: "He said that nothing is, even if something is, then it is unknowable."

Plato already asked the question: "Does man effectively benefit from autonomy of judgment or is he immersed in a virtual reality which cannot be differentiated?" In the dialogue of Theaetetus he makes Socrates say: "Are we sleeping awake and dreaming of what we are thinking or are we awakened and converse together?" And Theaetetus answers: "All is the same and accurately corresponds in the two cases."

This problem raises the question of the "evil demon" suggested by Descartes (we will come back to this point later), which still holds. Who can be sure of what, as far as the brain is concerned? If we are living in a virtual illusion of reality, what is to be copied if we try to identify our brain with a computer?

### Later On

Afterward, the various constituent parts of the Greek philosophy acquired the status of independent disciplines (natural sciences, physics, mathematics, medicine, psychology, etc.). Even the great philosophers of the 17th century, such as Descartes, Pascal, or Spinoza, made a clear distinction between their scientific work and the philosophical one.

---

[2] *On Nature or the Non-Being*, Gorgias of Leontini (483–375 BC).

Descartes compared philosophy to a tree "the roots of which are metaphysical, the trunk is physical and the branches are every other science, such as mechanics, medicine, morals. ..." Many of those philosophers have questioned, with only words, the brain mechanisms, the possible emergence of a new human being, or even the possible virtual reality which may already surround us. Some of them have left deep traces and cannot be discarded given their brilliant insight into deciphering the subtleties of the brain.

A recent re-edition of a text from Descartes[3] gives more details on the way to direct our mind and how to use all of the resources of intelligence, imagination, senses, and memory. This is an inventory of our means to know and applies clearly to mathematics. It is proposed that deduction proceeds from linked intuitions to bring us back to principles. The theory of knowledge is supported and guided by experience. But the "evil demon" is a shrewd and powerful bad demon. A misleading God who created us, such that we will always be fooled. This is the theory of the "hyperbolic doubt" seeking a primary truth: "I would think that the sky, the air, the earth, the colors, the figures, the sounds and all the external things we can see are only illusions and deceptions."

However despite his power, the evil demon has not the force to doubt his being, the certitude of his *cogito*. If the demon is misleading, he has himself to "be" in order to be misled.

This point was taken up later by Henry Putnam, in his fiction "The brain in a vat."[4]


## Nietzsche

Thus spoke Zarathustra: "What kind of man do we have to grow, to want, what kind may have the greatest value, will be the worthiest to live, the most certain of a future?" This question is very close to our research on the best brain to be copied. Nietzsche emphasized two types of men: the "last one" (the lower) and the "superhuman," whose nature is close to the

---

[3] *Regulae ad directionem ingenii: Texte critique établi par Giovanni Crapulli avec la version hollandaise du 17eme siècle,* Rene Descarte, Martinus Nijhoff, 1966.
[4] *Reason, Truth and History,* Hillary Putnam, Cambridge University Press, 1981.

divine. These are the two examples of what humanity could become and they call for the eschatological question of the very purpose of man, his capacity to possibly create a future. The last man is one of the figures of the "nihilistic disaster" which threatens the occidental culture.

And Zarathustra added: "Never before has there existed a superhuman. I saw both of them naked, the tallest and the smallest of the humans; they still look too similar. The truth is that the tallest seemed to me exceedingly too much human."

Of course, exactly understanding the thought of Nietzsche is not so easy. The question remains as to what to do with man from an idealistic point of view, if it could become possible to change him or make an improved copy.

## Spinoza and the knowledge

How to gain the knowledge, what is the role of perception, and what are the steps to follow? These are the topics of "On the improvement of the understanding." The first stage, says Spinoza, could be to learn by practice, which means elements the consciousness accepts when nothing contradicts; but they are not necessarily ascertained. The second stage relies on the "empirical perception," usually accepted to still be irrational. Then a deductive rational perception can arise from a logical procedure from an element considered as true and not result in a contradiction; finally, the intuitive knowledge is recognized as true (the knowledge of God is the absolutely essential starting point).

Spinoza, then, draws heavily on the Cartesian theory, as explained by Antonio Damasio,[5] a genuine Californian neuroscientist. Descartes was wrong and Spinoza was right: mind and body cannot be dissociated.

## Kant

Last but not least, Kant also asks the idealistic transcendental questions: "What can I know? What must I do? What am I permitted to hope for?"

---

[5] *Spinoza avait raison: joies et tristesse, le cerveau des émotions*, A. R. Damasio, JACOB, 2003.

Philosophy intimately mixes theory and practice but everything which belongs to the transcendental domain is beyond the reach of the human mind.

Nowadays, new fields emerge which still obey the generic term "philosophy" but with a scientifically assumed connotation. Bioethics is devoted to neuroscience, the progress of which allows, gradually, reading or even modifying the brain's activity. The "language philosophy" seeks to establish relationships between language and thought. The mind philosophy raises the question of the functioning of cognition.

In the 20th century, the analytical philosophy as represented by Grege, Schlick, and Russel, among others, took over the philosophical classical principles with the aim of clarifying by means of intensive use of mathematical logic and critical analysis of the wording. In all of that, we are no doubt in a process of decoding the programming of the mind's mechanisms.

Philosophy constitutes a critical approach, in a constructive meaning of the term, to bring to light the illusions or the contradictions. Descartes became famous for his theory of doubt, leading to rigorous thought. The intended goal always remains to understand the reality in a deductive way, to distinguish the objects by raising ambiguities; formal logic is therefore involved. One implicitly seeks to emulate the mathematical rigor (Spinoza, Leibniz), but without the help, in the cognitive field, of symbolism.

Here are just some intellectual acrobatics that a philosopher's brain is able to perform to introspect without any instrument. The machine, of course, will have to take them into account.

Animals actually also enjoy a kind of intelligence (look at your cat, for example!) directly oriented to the survival requirements. What makes a definitive difference from human beings is that they are not as able to speak in an elaborate language as humans. Thinking, as well as memory, requires words to proceed; even images, sounds, and smells induce words in the background.

## *Psychology, Psychoanalysis, and Psychiatry*

These three fields of research emerged directly from philosophy, with the aim of bringing some scientific methodology and a tentative experimental approach.

## Psychology

Psychology was directly inspired by philosophers who tried to get logical explanations about the mental processes they were acquainted with, such as the structure and the functioning underlying psychism. This involves an *a priori* and intuitive knowledge of the feelings, ideas, behaviors, and ways of thinking.

Of course, that requires, first of all, a clear basis of this evanescent notion we call "normality," and that is not so easy. Could we, scientifically, rely on the cold statistics? How to take into account the environmental considerations? Could we analyze the mental structure borrowing from software examples?

All of that brings us closer to the logic and the heuristics of the mathematics. Norbert Wiener, the famous theorist of communication theory, was the first to build up a comparison between the brain and the computer. John von Neumann, the physicist of quantum theory and game theory, was a pioneer in artificial intelligence. Alan Turing asked this fundamental question: "Would a machine be able to think?"

However, should psychology be considered a real science or a non-rational subject? What could be the relevant level for a psychological analysis — neuron, network, or neurophysiological mechanisms? Of course, the modern means (EEG, MRI, and so on) do contribute in correlating the cognitive psychology with the cerebral localization to achieve an informational modeling.

Many "philosophies" have emerged to answer these questions of how mental states occur. Functional psychology (or functionalism) was initiated by William James in the late 19th century; it is not based on experiments and is thus criticized by "behaviorists" or "physicalists." All of them try to identify the role of mental and physical substances, which Descartes took as independent. Currently, the background idea of these countless theories, even if not always clearly stated, remains as a comparison to the way software is managing computers.

## Psychoanalysis

Psychoanalysis was introduced by Sigmund Freud (a neurologist whose theories considerably impacted 20th century philosophers). It aims at discovering the mechanisms underlying the operation of the brain of every person — mechanisms which guide their behavior (psychism); also, this effort should lead to the deduction of a theoretical, global, and universal model.

The starting point is that the subconscious states have to be primarily considered because the subconscious is well known to rule consciousness. This implicitly suggests that a psychological determinism may govern the mind. Every idea leading to an act is not arbitrary but arises from an induction mechanism, a preceding activity which could be revealed by an exploration of the subconscious as long as a reliable method is available (dreams, sleep, errors, unconscious omissions, etc...). The idea, before reaching the consciousness, suffers some changes but roughly remains linked to the initial idea; one may try deducing a method of interpretation of these influences.

Obviously such descriptions of psychism, qualified as "scientific," may be subject to a criticism inseparable from the Freud personality: cathartic method, therapeutic recollection, etc. The reply has sometimes taken the form of a strong critique of the philosophers: "The philosophers are similar to this man walking in the middle of the night, with his night-cap bonnet and candle, trying to fill the holes of the universe" (Heinrich Heine[6]).

## Psychiatry

Apart from the exploratory disciplines we have just discussed there is psychiatry, which is intended to discover from the behavior of a person possible mental disorders, mentally deviant behaviors (possibly patho-logical), and provide an adapted therapy. These psychiatric illnesses can be divided into various categories according to their outbreaks: psychoses, schizophrenia, anxiety, phobias, nervosas, and feeble-mindedness. All of these attempts at categorization have inherent fuzzy limitations and could potentially be influenced by addictions (alcohol, sex, and drugs).

The central problem of psychiatry remains in the search for a scientific and accurate definition of what is called "normality" and its

---

[6] *The Return* (poem), Heinrich Heine.

limitations. Would we have to start from a statistical mean value established on the basis of fuzzy criteria to be defined? Does this mean value have to change with the level of instruction or education? How to quantify these notions? Is this limitation on the normality a rigid barrier or would it remain vague, knowing that the individuals and their brains are eminently variable and diverse? Could we reasonably make a comparison between the relative merits of a pear and an apple, which both are anyway delightful fruits?

However, psychiatrists (now renamed "neuropsychiatrists") nevertheless identify mental health diagnostics. Are they reliable, and could they be used in the description of a brain? Psychiatrists also propose therapies ranging from neuroleptic or psychotropic drugs to electroshock, or even more extreme solutions, such as lobotomy.[7]

Normality tends to classify brains into two distinct categories: normal and abnormal. Would the latter have to be rejected outright in our attempt to build a software model of the ideal brain? This is very uncertain, because we would be trying to reject some "abnormal brains" which, however, succeeded in demonstrating prodigious intelligence: Einstein could have been regarded as abnormal as he did not obey the norm; Picasso and Dali were not insane — contrary to van Gogh, who also produced masterpieces and yet was not so far from madness. Would we have, as Erasmus said, to evoke a "praise of folly" in order to recognize exceptional brains?

What is then to be learnt from all of that, if the final intent is to build the best artificial brain and, above all, if this brain could be "improved" and updated? Improved with respect to what? Would we then have to inject into it a little grain of folly? Would we have to psychoanalyze the computer so as to be sure it does not create nonsense when it behaves, in an "exceptional" manner and is then difficult to evaluate? Computers and their Artificial Intelligence already do raise such issues.

---

[7]Lobotomy is a major surgery (now quasi-universally forbidden) which consists in sectioning some nervous fibers in the white substance of the brain. This operation is especially devastating and is practiced in a totally blind manner; it often results in irreversible damage.

# The Complexity of the Information Processing

## Neuronal Exchanges — Is It All in the Genome?

A gene is a stretch of DNA that is devoted to the making of a particular protein molecule. Some genes are active (a third of the 20,000 genes that make up a genome, mainly in the brain); others not (as far as we know today). An example is the ASPM gene, which makes a protein directly responsible for producing new neurons; others are involved in the connections linking the neurons together in a network, and still others command the information transmission (neurotransmitters). That is to say, that gene is of key importance in the brain's activity, through this ability to grow proteins.

When one adds that the brain is the only place in the body where the genome differs in each of the neurons, it follows that defining the exact role of the genes in the brain functioning (and diseases) will be of huge complexity. Craig Venter, for his part, had the opinion that "all is in the genome." That may be somewhat excessive, but the contribution is certainly essential.

Single gene mutations occurring in the neurons might be responsible for neurological disorders such as Alzheimer's disease, strokes, or autism, but things are not so clear to date.

## Predeterminism, Randomness, and Necessity

A key question that is still unanswered: What the heck exactly does "to understand" mean? Not so easy to get a clear definition! We could say that there is a procedure for assembling all the information available in the memory on a definite subject and making a critical assembly of these elements; if all the stuff fits well, without disturbing contradictions, we could say "I understand" (and careful people would add "to the limit of my knowledge"). The mind is trained for such a drill; if the result is not convincing or something is missing from the puzzle, we say "I don't understand" or "Something is wrong." This could be considered as a pure software operation which can be mimicked on a computer.

Intelligent people rely on a deep knowledge — immersed in the memory and the unconscious — to reveal all the possible correlations between the elements and reach a sound judgment. Making a comparison with the computer situation, we would say that the data mine is "big." Otherwise the guy is qualified as stupid!

This is a critical approach driven by the consciousness and the subconscious. The computer, to date, is only able to dig in the memory and reveal correlations but has no access to the final step in the consciousness which "appreciates" the soundness of the data.

Google and its search engines are very swift at hunting down the key words in a mountain of references but there is not yet any intelligence behind them (for the moment, I stress). There is a hierarchy of difficulties to overcome.

This is close to the Turing test, which decides when a computer can be treated as intelligent: the test brings together a man and a computer (that have never met before), through an anonymous phone call. Would the man be able to tell that he is talking to a machine? This test might have been fulfilled with 12 English speakers and a computer which impersonated a young Ukrainian guy. A majority of the speakers were duped but the involved field of intelligence was rather narrow. This is the very beginning.

The knowledge of the biological functioning of the neurons would have to be associated with the knowledge of the functioning of the whole organization and the synchronization of these billions of cells in order to "understand." It is clear that imitating something we do not understand would *a priori* be fatally flawed.

Jeff Hawkins, an American neuroscientist, laid out[8] his "memory prediction framework." According to him, the brain is organized in a hierarchical mechanism which in some way "predicts" the future input sequence, thus memorizing the frequent events or patterns which are so ready to be used by cognitive modules in a possible future similar situation. These "anticipatory cells" are distributed in the whole cortex and can be fired before the actual object emerges on the scene. If the operation is judged irrelevant, the process stops before reaching the consciousness. This means that the brain is able to store and retrieve information, in order to intelligently foresee what can happen. What looks very surprising in this biological process is the quickness of the treatment of the

---

[8] *On Intelligence*, Jeff Hawkins and Sandra Blakeslee, Times Books, 2004.

information, as long as it does not refer to an electronic device but a biochemical one.

Would all of this recent knowledge of the brain mechanisms of information processing lead to a "thinking computer"? The key difficulty remains to give it autonomy of judgment through the implementation of a "consciousness" (I hesitate to say "a self").

## The Role of the Five Senses

The brain's activity is globally rather confused and messy, but permanent. The only signals to survive are collective, organized, or corresponding to a set of coordinated neurons. The current methods for detecting the brain's activity (CAT, MRIf, etc.) only bring information on the consequences of the activity, not on the origins, the motivations, or the way they are processed. Those are only signatures of activity markers. The five senses are there to bring a diversified and understandable contact with the environment. Moreover, they are able to set off an alarm when things go wrong: this is pain, whose role is to help prevent greater damage to the body.

Very thin electrical or optical probes make it possible now to detect the activity of quasi-individual neurons, the problem still being to make a motivated choice among the multitude of neurons and address specifically the one that has been chosen.

### The Origin of the Exchanges

The brain's dominant source of information lies in the five senses and, there also, unexpected behaviors are observed. First of all, there is what is called the "cocktail party effect." Obviously, the sound information comes from the ears, and then the corresponding signal heads towards the brain, where it is analyzed to give the sensation of sound.

This is clear but difficulties arise when the brain has to identify the words individually in order to "understand" the meaning they are related to. A new discipline called "psychoacoustics" has arisen, and it is devoted to decrypting this process in order to draw inspiration to make a copy in a computer for speech recognition applications.

Surprising breakthroughs have been made which confirm the idea that the brain is undoubtedly a marvelous machine: When two persons are speaking simultaneously, the brain can — to some extent, and by focusing — follow selectively one speaker whilst remaining subconsciously vigilant in the other "channel," and do this in real time. This performance can in no way be performed with analogic electronics, and even the use of a computer is not straightforward because the two messages are physically similar and cannot be separated or filtered as a noise would be. The passbands are identical. It can be supposed that the brain, for its part, is able to get and separate the words in both channels by comparing the imprints with what it has in its memory, and then rearrange them in separate queues we could follow selectively with some mental effort. All of this work remains at the subconscious level before reaching consciousness when achieved. This processing has to be fast enough in order not to delay the information which is to keep in real-time.

A similar operation is performed with vision, which is able to reconstruct a single and coherent (3D) image in the brain from a multitude of independent signals originating from two different retinas differently oriented. The idea we have of the external world is a purely virtual construction of our mind. Do we have to trust it? What is the limit of this illusion? Magicians empirically know how to manipulate the vision of what we take as the real world.

## Words in the Brain

Words are the drastic difference between the human brain and animals. Young babies, at the same time as they learn how to breathe, become familiar with the sound of the words. Some people even say they have been previously sensitized to the sounds when still in the womb. The babies' first intellectual training will be to learn how to speak; this slow learning progress with the help of the parents (mainly the mother, if she has time enough to participate[9]). The whole activity is devoted to deeply building the mind, the memory is entirely devoted to getting the basic

---

[9]Modern life does not favor this essential step for the baby who dramatically needs permanent and affectionate contact with his mother in his first learning of life. The same thing is true for any mammal or bird.

words, and there is no place for remembering the events of life until the age of four or five. This is the time when the basic building of the brain takes place; subsequently it will be more difficult to alter the mind — the neurons have already taken the trick.

This is the time to learn the words and their organization (grammar), and this is for all life. The brain is open to this learning of a language[10] as neurons are proliferating and organizing; things will change dramatically with advancing age, and that is the reason why learning a foreign language is much easier at a young age than later, when the learning mechanism has turned to other targets.

Words, their meaning, and their way of being used are essential for the following, because we think with words. The richer we are in words, the more adapted we will be to "think"; memory, then, will do the rest to build up intelligence, but words are the inescapable basis. To create an intelligent mind requires careful training in the language, possibly assisted by complementary instruction in an ancient language (Latin or Greek, to achieve a full knowledge of the meaning of words), which will help the brain to obtain the exact meaning of words through etymology and prevent confusion.

The computer has no difficulty in obtaining words or even grammar — but how to explain to it what that really means? That is the problem in making a computer intelligent; it does not know the meaning of what it learns — it does not "think" by itself.

Neural networks aim at mimicking the neuron organization in a "supervised" or even "unsupervised" operation, closer to how the brain learns. So Yann LeCun (from Facebook) made a decisive contribution to improving the "backpropagation" in the "hierarchical construction of the characteristics." Assuredly he does know what it is, but he nevertheless said: "How this is done in the brain is pretty much completely unknown. Synapses adjust themselves, but we don't have a clear picture of what the cortex algorithm is. The inner working is still a mystery."

---

[10]This is the reason why primary education should be exclusively and completely devoted to the three Rs: reading, writing, and arithmetic; nothing else. The students will have, later, all the time in the world to learn how to use their smartphones!

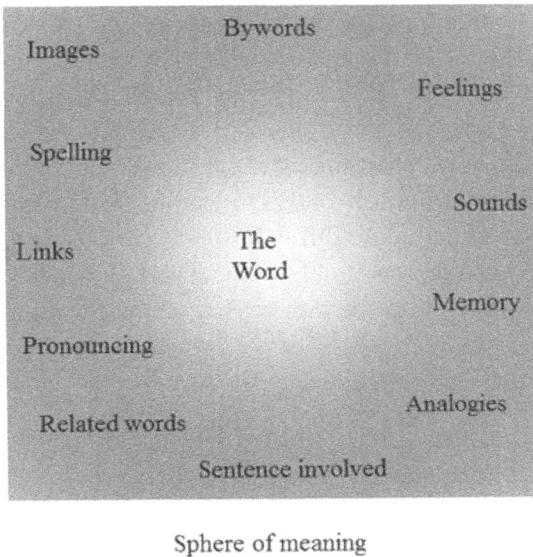

Sphere of meaning

**Figure 2.1.**   Sphere of meaning.

In spite of that, impressive achievements flock together in the current news: the recent success of the AlphaGo algorithm in Go competition takes us closer to mastery of a rising intuitive behavior of the machine.

Words are footprints of the thought alchemy. Focused attention and thinking requires classified words from the internal world of the unconscious. Words can be directly descriptive, like images or feelings, but they can also be "intuitive" or unrelated to a specific object (sentiments, opinions, moods, etc.); in any case, words come with relationships, to varying extents, with other words to complement their meaning as an indivisible whole. This could be considered as a "sphere of meaning" that goes with them. This is tentatively sketched in Figure 2.1.

Of course, such paraphernalia largely depends on the language used, and this is the reason why translating a word into its claimed equivalent always remains an approximation in the sentence. If a "thinking machine" is to be elaborated, it should mandatorily take into account the constitution of such spheres of meaning for each recognized word. This is already especially valid for translation software.

There is also an important mechanism in the *modus operandi* of the brain: the oblivion. Many recorded elements are kept in a closed part of the brain if they are to be reused in a near future. The information has to be categorized as a function of a possible usefulness. As long as this use is not recalled in mind soon enough, the "rank" of this information is stepped back and progressively enters in the oblivion domain from where it will become more and more difficult to retrieve it. The brain then tries (as a computer does) to reassemble immediate key words or other information known to be next to the target. This allows recalling the vanished memory. This careful rearranging of the memory represents a permanent activity of the subconscious, day and night.

## The Second States

So-called second states correspond to situations where the brain is temporarily disengaged from reality, but that does not mean that it is inactive. A relation still exists with the real world, if only to maintain vigilance.

### Sleep

Sleep is as essential for the brain as wiping and defragmenting are for a hard disk (such similitude is not unusual)! The difference lies in the fact that the brain continues to work during sleep even if consciousness remains in the background. The search engine of the memory classifies, rearranges, and evaluates the reminiscences, the remembering, and the analogies to be used later or recalled when awake, without any conscious rules or the trigger of a mnemonic.

This night operation of rearrangement of neuronal circuits contributes to the strengthening of knowledge, and reveals correlations or significant analogies. It is not uncommon, on awakening, in a half-sleep, to see a clear solution to a particular concern that we were not able to resolve the day before.[11] Then we need to make a voluntary effort to memorize it or write it down immediately on a piece of paper, because the idea might be

---

[11] The story goes that Archimedes was in his bathtub when he shouted: "Eureka!" This is not certain — he might have been awakening from his bed!

left again in the subconscious, from which it would be difficult to make it re-emerge later. This transitory re-memorization is known to take place in the prefrontal cortex of the brain.

Other circumstances may help one to communicate with the subconscious: meditation, Zen, or religion, which maintains contact with the transcendental or the spiritual; sounds like those of an organ in a cathedral also help to relax the brain. The silence favors non-thoughtful solutions. We will deal later with the modern theory of "neurotheology," which aims at finding a neurologic origin to the idea of religions.

However, sleep mechanisms remain a mystery, which some are trying to explore.[12] Wearable sensors currently allow the gathering of data about the way sleep occurs and evolves. Data are accumulated from many volunteers, who describe and correlate the electrical activity, the movements of the body, the breath, or the heartbeat to make an appraisal of the "quality" of the sleep. We will deal later with this new field of investigations in the framework of big data mining.

## Hypnosis

Hypnosis was also considered, as sleep, to modify the brain's functioning. For a long time, this strange behavior aroused scientific curiosity but the experimental means of investigation were rather poor and no conclusion appeared available. Things changed with the modern arsenal of PET, MRIf, and EEG. These studies showed that hypnosis cannot be confused with sleep; the phenomenon is different and involves a network of cerebral regions in a collective activity, whereas other regions, usually very active, are shut down. The way this phenomenon is managed is still to be discovered (cerebral waves, sounds, images of the scene?). What exactly is the corresponding level of consciousness, and would a computer be able to simulate (or stimulate) hypnosis? There are so many questions to be answered before one gets a satisfying explanation!

All these analyses contribute to explaining the brain's operations, such as pain perception or the interactions between cognition, perception, and emotion. However, Pandora's Box has just been opened and no

---

[12] For more details, see for instance *The Tides of Mind*, David Gelernter, Liveright, 2016.

definitive model has been proposed for such a reconfiguration of the communication between the concerned regions. Neither has a conclusive explanation been reached, for instance to describe the triggering of hypnosis.

So what could be proposed for the computer? Would it also have to be hypnotizable in order to fit exactly with the human brain?

## The Experimental Tests

We have already seen that physical means are now available which help investigate the geography of the brain's activity: EEG detects the potentials evoked by the movement of electrical charges inside the neurons or from one neuron to another; MEG detects the magnetic component of these currents. Both of these properties can be reversed to some extent, to induce an internal localized electrical action as long as the zone concerned has been identified. MRI and PET provide images of the activity in the brain; nevertheless, there is no means of knowing the basic reason for the firing of a neuron or the origin of the collective organization.

To find out more, intracranial recordings are also used by electrophysiologists on patients with established neurological disorders. Thin electrodes are placed (under a CAT control) to reach a specific zone of neurons or possibly even a single neuron (inevitably a bit randomly, because an individual and motivated choice is hardly imaginable). Such a method is, of course, particularly sensitive and localized because it is very close to the emitting neuron.

Each neuron being specific to a particular function, it becomes possible, by elimination, to roughly know what the patient could be thinking of. It should be difficult to choose the "thinking" but this could be reactivated by the memory without reaching the field of consciousness. The conscious information is concerned not with a single neuron but with a multitude of neurons which are collaborating in the "thinking."

Of course, such experiments on humans are necessarily limited to specific cases where nothing else can be proposed, but animals have no such limitation and are more extensively studied (with the inherent difficulty that we definitively ignore what they "think").

Coming back to humans, one may also extrapolate this external stimulation with the aim of artificially creating and imposing a particular "state of awareness" by exciting the corresponding regions of the brain (for example to restore the memory of an event embedded in the subconscious). This is, in some way, what happens spontaneously in the case of hallucinations or even dreams.

An extension of this idea was some time ago proposed by the philosopher Hilary Putnam with his famous imaginary experiment: "the brain in a vat" (a modern version of the Descartes's "evil demon"): he imagined a brain put in a vat and provided with every fluid necessary to keep it alive biologically. Then its consciousness could be maintained in activity by external electrical stimulations originating in a specially programmed computer. Would this brain be conscious that it is a toy of a somewhat mad scientist or would it take its feeling as the "real world"? One could then study quite easily its spontaneous reactions and its artificial and conscious feelings, free of any external influence.

But the brain, to ensure it operates correctly, requires it to be in symbiosis with a body. It has to be accompanied not only by an oxygen supply but also by various molecular influences: adrenaline, hormones, cellular waste, etc. To "isolate" a brain, besides the necessary nervous links, delicate and complex biochemical backing is required.

Many other questions are still pending. It also remains for us to understand how a voluntary thought could be "externalized" and deciphered. Would the brain be limited to "functionalities"? Could we imagine making a brain able to survive after the death of the body?

We have identified correspondences between the cerebral activity and the mental life but there remain abysses to be filled.

In the cortex there are about 16 billion cells, each of them transmitting a few bits of information. This crowd of neurons occupies a rather large surface; that is the reason why the brain is "ruffled" and not "flat," thereby providing an extended available surface in a limited volume.

Most of the "intelligent" activity is located in the frontal cortex, and this is why intelligent people are expected to have large frontal cortexes — such as Professor Nimbus! However, animals also have a brain and they do use it. The animals which are "pack hunters" are known to have a developed frontal cortex; an example is lionesses, which use elaborate and coordinated group strategies to reach their prey; conversely, lions, which do not care about running after their prey, have a less developed frontal cortex.

In the building of information in the brain, the neurons which are inactive are as important as the others, because they indicate what is not contributing. Basically, in principle, there is nothing wrong with simulating the brain's activity through a software program as long as one is able to conveniently analyze its complexity.

Some pathways of thought are currently being investigated and the corresponding external signals have already been identified. It then becomes possible to easily implement the control of devices by thought (articulated prostheses, vehicles, any connected object, etc.) through the electrical signals collected externally and directed to the computer.

However, we are still a long way from being able to understand and to explain the everyday experience of conscious life. This will certainly be a task not easily translated by a computer program.

## Computational Neuroscience

This field is the study of brain functions in terms of the information processing.[13] It is a highly interdisciplinary field which gathers together neuroscience, cognitive science, psychology, and also more "scientific" classical sciences (electronics, computer science, mathematics, and physics).

The aim is to translate the biological behavior of the nervous system into electrical interactions and model the processes into a computer language.

The initial work was devoted to the "fire" model of the neuron and the buildup of the action potential. Then the cortical areas, which process the information from the retina, were investigated. Currently, the focus is very much on computational modeling of single neurons or dendrites and their complex biophysical behavior. Also, more global approaches are under study, such as the axonal patterning and corresponding sensory processing toward memory, cognition, and consciousness. This creates a particularly extended field still to be explored.

---

[13] *Computational Neuroscience,* Eric Schwartz, MIT Press, 1990.

Then the brain's activity cannot be dissociated from the notion of intelligence. This special feature of the human brain (?) directly relies on the relation which man has with the external world through the sensations provided by his body and the "experience" it provides and accumulates of the surrounding physical reality.

There are people for whom the relation with the world is rather limited (for instance Stephen Hawking when alive), which does not prevent them from being outstanding intellectuals. Abstraction is their domain. Seeing and hearing are nevertheless critical for the acquisition of the knowledge as well as the physical means to express oneself. But a better grasp of the external world is a *sine qua non* condition for the development of the brain.

The human brain seems to have an enormous capacity for memorizing and recognizing images, among them human faces. It will be of primary interest to decode the involved mechanism in order to potentially replicate it in an intelligent computer in order to make it able to decipher the facial expressions. This analysis is now in a booming progress and is already used in a wide range of applications.

Then intelligence proceeds from observations and simple images: recognizing that a screw and a bolt are made to be assembled is a step of intelligence; this can be taught to a chimpanzee, but if the bolt is changed, the chimpanzee will never be able to understand why it is not working anymore. An intelligent machine will have to be able to sort and associate data but also to face unexpected circumstances (as the chimpanzee would have).

In the more delirious speculations of physics or mathematics theories, it remains essential to refer to well-known analogies, words, or images of the real world in ways that peers can understand and follow the development of the ideas (quanta, string theory, tunnel effect, dark matter, sets or rings, waves, etc.). An intuitive material support is mandatory. We cannot escape the reality of our material environment. Ideas are made of words, and words have precise meaning.

This implies that, in order for the machine to duly become intelligent, it must fulfill three basic requirements:

(i) Be provided with senses (detectors, sensors, etc.) equivalent to (or better performing than) the human ones.

(ii)  Be piloted by advanced software capable of learning and memorizing.
(iii)  Tentatively be able to "think" — that is to say, have a consciousness which makes it able to make a reliable evaluation of the situation.

Of course, the third requirement is the key one, the most difficult, and is presently not satisfactorily fulfilled.

Consciousness in itself remains a deep individual mystery close to existential beliefs and religions; many research programs are devoted to it all over the world, mixing many different disciplines. To some extent, is this to be considered as the final grail?

We will deal with it in the next chapter.

# Chapter 3

# Be Aware of Consciousness

This topic is especially touchy and fuzzy in its very conception and it is not so easy to establish a non-ambiguous definition of the field involved. Several words pile up to make sense of it, each of them inaccurate; this is in no way surprising, because it is about the intimate relation between our inner personal world and the external one.

These words compete:

- *Awareness* refers more specifically to the way we appreciate or know something; it is close to "knowledge."
- *Conscience* preferentially leads to our morals or our evaluative judgment. We often say "have a good conscience."
- *Consciousness* is more general and relates to the perception we have of the external world through our senses. It is in no way related to a notion of value; just a perception, a description of the details as they come to mind.

I guess the third item is the closest to our purpose, and we will keep the word "consciousness" in the following.

## Observed vs Observer

What we are dealing with looks like a kind of iceberg. We begin to get an idea of what is in the part above water: consciousness. But the most important thing is to be discovered and it is submerged below the surface of an ocean of ignorance: the unconscious, which manages everything in our brain.

### Consciousness

Consciousness can be considered as the *forum* of ancient cities where ideas meet, exchange, compete, and interbreed, coming from the subconscious, perpetuating for a moment, or disappearing without any apparent reason, the public space between the external worlds represented through our sensations and our bubbling and uncontrollable inner world.

From Plato to Descartes, the philosophers,[1] the poets, and the psychologists have vainly attempted for a long time to understand consciousness — that is to say, drawing limits or models to get a description of this virtual reality which is familiar to all of us but remains of inextricable complexity.

We will deal in a few pages with the issue that "psycho-cognitivists"[2] analyze in stacks of books, entering into the details and relying on extremely varied tests.

What exactly is the purpose which consciousness serves? Alexandre Vialatte, a French author, jokingly wrote: "Consciousness, as the appendix, is of no use, except to make you sick." Is it, even so, an illusion, a mystification, an unnecessary observer of the events that are beyond it? Do our decisions come from the deep subconscious? Consciousness could be the referee before making decisions, resulting from a long learning, separating the possible from the plausible, given the subconscious context.

Then the following question arises: Could a computer-compatible model of this (undefined) concept of the consciousness be conceivable? Would it be perfectly faithful to the original? Could it be improved or updated so as to make it more effective, more efficient? Whatever care we take, that computer will necessarily represent a different person even if very similar. Such a robotic situation has been emphasized by many

---

[1] Putnam, *Reason, Truth and History.*
[2] *Le code de la conscience*, Stanislas Dehaene, JACOB, 2014.

authors, among them Hans Moravec,[3] with "his mind children." Would that be the divergent point between man and machines?

Finally, I have a question of my own: I read carefully the excellent book *Le code de la conscience*[4] (*The Consciousness Code*) by Stanislas Dehaene, exclusively devoted to consciousness, and I was very surprised to never meet the word "intelligence" in the discussion. Would these concepts have to be separated? I guess not.

Consciousness is the very prerequisite for intelligence which can be considered as an ability to bring to the conscience, from the deep mind, the most convenient elements to contribute to the present idea. It obviously constitutes a platform for the evaluation of the ideas, whereas getting new ideas is the very nature of intelligent minds. So these two concepts remain inextricably linked together. One can, to some extent, be conscious without being "intelligent," but never the opposite.

This is the reason why, if artificial intelligence is to be comparable to its human equivalent, it should necessarily include a dedicated software module playing the role of consciousness in the elaboration of intelligence. As a matter of fact, the automations we are familiar with (such as a thermostat or any regulator) already represent an elementary kind of consciousness limited to the only parameter concerned. There is nothing, in principle, which opposes extending the domain of their consciousness at the same time as their intelligence determines the action (Google car). But how to make it conscious of such a consciousness?

After assimilating and probabilizing the concerned elements taken from the subconscious and the data from the body, consciousness brings them together into a synthesis, a logical evaluation of them. This process may lead to a conscious decision or, in case of emergency, directly induce deliberate actions.

Such a bypass can even be especially developed in sequences where consciousness plays a rather limited, *a posteriori* role of simple monitoring, for instance in the case of a pianist who does not command his movement for each note but plays instinctively. The role of consciousness is to

---

[3] *Mind Children: The Future of Robot and Human Intelligence*, Hans Moravec, Harvard University Press, 1988.
[4] *Op. cit.*

create and organize steady thoughts which will be memorized. Would a computer be able to do that with the same flexibility; would a computer be able to establish an independent thought?

We do know that random neuronal fluctuations (neuronal noise) are permanent and underlie incongruous thoughts which have to pass through the criticism filter and are allowed follow-on only if validated. There is no divine operation in "inspiration"; the only chance orders the emergence of the flux of these images or these words which fuel our mental life.

The unconscious instinctively, spontaneously, and in a concealed way uses the basic properties of the conditional probabilities (the famous Bayes equation) to reach weighted statistics in the framework of what could be a "fuzzy logic."

In a human brain, an army of bioprocessors, at any moment, assesses every option of the conscious field and everything which could be of some influence; but the unconscious presence of the conscious referee remains necessary for guiding to the decision (as a BTW[5] assistant). When a person is anesthetized, the process gets blocked.

The unconscious is full of unrealistic information we never recognize. The US Army has an ongoing active research program whose aim is to browse in the subconscious to reveal information which has been perceived but remains stored, somewhere unreachable, in the brain. Images or words are scrolled in a computer-assisted operation to generate unconscious reactions to these stimuli and reveal valuable information embedded in the depth of the mind. Man then plays only the simple role of a passive and unconscious detector.

The unconscious runs "in the background," like a computer, to manage a huge volume of data (quite binary) which strongly resembles the big data of the databanks we find on the Internet when rummaging with an ingenious search engine.

## Introspection

Besides, the question can be asked of how far such an introspective approach makes sense and whether it is not, by nature, doomed to failure.

---

[5] The "bearing to waypoint" of airplane navigation!

Indeed, the "object" of our observation is still an integral part of the observer! In such an auto-analysis situation, can it be reasonably expected that we can reach unquestionable conclusions? How can we be sure we have stayed within the limits of a convincing impartiality?

The blindfolded man analogy we used in Chapter 1 embodies here its whole dimension, although the supermarket is not integrated into the brain of the blind man, which would certainly complicate his work. Maybe this could be experienced with a "virtual supermarket"?

Anyway, this seeking for selfhood was not born yesterday; it was the main driving force of Greek philosophy at a time when science and philosophy were one. Rationality and logic had just begun to become accepted as a way to think of and analyze our own brain. Nevertheless, opponents of this philosophy, such as L. Weiskrantz,[6] defend the idea that introspection is not to be trusted.

Consciousness is the key element of our mind, our self, maybe our soul too but the way our brain manages the consciousness is still a deep mystery. Some argued about the dualism of the body and the soul which seem to both be involved in our consciousness. Plato thought they are separate entities, Descartes for his own part imagined that the body, including the brain, is a mechanic which has to be analyzed as a machine, but the soul (and the language) is elsewhere, and knowledge about it is inaccessible.

Consciousness allows sequencing of logical and rational ideas following schemes that were acquired by education. Nevertheless, we are at the limit of a possible unconscious anticipation, a limit which can be overridden in the case of unpredictable instinctive movement.

A particular point refers to mental arithmetic: one is conscious of the sequential strategy of the calculation; one can even choose the method to follow, which is not usual in psychology. An introspection step by step of the algorithm is even possible while keeping the provisional results in mind. The brain works as a hybrid serial/parallel machine which requires ongoing monitoring of sequential operations with a temporary storage memory (as in a pocket calculator).

---

[6] *Blindsight: A Case Study and Its Implications*, L. Weiskrantz, Oxford Clarendon Press, 1986.

Nevertheless, it could happen that very complex calculations are performed in a purely instinctive and subconscious way. This process could, however, be dominated by the subconscious when the sequence has been memorized as a reflex: the "gifted calculators" do not need to "do the arithmetic" — the exact result spontaneously emerges.

So did Shakuntala Devi,[7] a famous Indian calculating prodigy, who said: "It just comes out. I don't have time to think about the process." To a smaller extent this may be experienced by anybody (not necessarily a genuine prodigy) when doing repetitive operations such as an addition of similar numbers;[8] after more or less extensive training, just seeing the list of notes and the result comes to mind without "doing the math."

Of course, this process of calculation belongs to a logic which the computer is fond of. It has been the most straightforward application of the machine.

Consciousness is the key element for the reconstitution of a brain; the machine that would copy the consciousness could in the first instance (to be trained or checked) only work with some reference information (internal, memorized, well-known). That should provide us with a model of introspection which could be helpful in a transfer to the human. In any case, before creating a simulation we need to know precisely what is to be simulated, and then it is the first requirement to investigate and dissect the underlying mechanisms and get realistic, logical schemes.

Thinking constantly modifies the organization of the brain, and that is what provides it with this unmatched plasticity. There is a specific biology of the consciousness. No two brains are identical, even when we are dealing with twins. Human brains perform very differently, depending on the actual conditions which prevailed during childhood, particularly the training in an elaborate language and the "handling" of ideas (Greek and Latin, which are nowadays forgotten in favor of our current technological languages but which represented a cultural, intellectual training very educational for the reflection and the consciousness).

---

[7] "Speed of information processing in a calculating prodigy", Jensen R. Arthur, *Intelligence*, 14(3), 259–274, 1990.

[8] For instance, adding up marks after an examination. Of course, checking the result is recommended (possibly with a more reliable pocket calculator!).

With their spellings and their corresponding phonetics, words and languages, even in their inexpressive form, constitute a largely secured message. The redundancy is so large that many of the letters or sylla-bles can be omitted without losing the underlying information which the brain can accurately reconstruct without any difficulty (telegraphic language).

## *Neuropsychology*

Another way to approach the issue is to step back from introspection which mixes the roles of the observed and the observer, and turn to inde-pendent analysis of the brain of the observed. That is the way psycholo-gists work to determine how a "normal" consciousness proceeds but also how eccentric affirmations can be explained rationally.

The brain is able to perform miracles in information processing, but it cannot do everything. What it is asked to do must be compatible with its software — that is to say, its previous learning accumulated from years of training, education, and storing tasks. Sight, speech, and hearing result from long training of the neuron organization.

But the brain education can also be bypassed by organized illusions. The underlying programming can be dubbed by some casual or deliberate gimmicks having access to the internal mechanisms. These tools, still primitive and rough, can possibly be used to manipulate the consciousness in order to understand how that works.

These mental illusions can arise even in unperturbed minds. A famous example is what is called the still-unexplained sensation of "disembodi-ment" that gave rise to so many passionate discussions and experiments in the past. Such a mental state has been artificially stimulated in some patients and it was discovered by EEG that its location in the brain was lying in the temporo-parietal region. Then, the invoked parapsychological explanation does not hold up anymore.

These sorts of illusions or hallucinations induce important difficulties in solving the problem of the fairness of the observer. Then, would the conclusions we may reach about it be meaningless? Would the "observed" be in a permanent and unreliable moving state which could prevent any consistent description?

In such a query, an issue which also remains is to be able to identify the cerebral bases without being "contaminated" by the observed. Then, the only experimental physical analyses (such as MRIf) allow an unbiased approach through indisputable measurements. Qualitative sensations, in some cases, can be so quantified in a measured parameter. In this field it has been noted that reliable and reproducible markers were empirically discovered which are typical of the process of accessing consciousness.

Moreover, another hardly explainable, difficult-to-define specialty of the consciousness is the establishment of what we call "the mood"; this fuzzy mental state could be good, bad, or simply changing, and lead to an *a priori* evaluation of the reality independent of any rational criterion. This is also related to our personality, our "nature." In order to mimic the brain in detail, would the computer also have to be dependent on an uncontrollable, unpredictable, and modeled "mood"?

Not to make things simpler, the overall functioning of the brain also implies several peripheral organs which are connected to the cortex and busily contribute to regulating the consciousness activity:

- The two thalami contribute deeply to the alertness, the attention, and the synchronization of the whole system. They are implied in the disconnection of the cortex with the sensorial stimulations during sleep.
- The hypothalamus, a small gland located just below the two thalami, is more concerned with metabolic processes and the nervous system (hormonal control).
- The hippocampus interacts in several processes: attention and spatial memory; it is also directly involved in the production of new neurons. It supervises the latent memory and the storage of strong memories.
- The basal ganglia, which proceed to the decision and action processes.

All of this cerebral organization which contributes to the consciousness is much more developed and structured in the human brain than in those of animals — even the primates. The relationship with intelligence is straightforward even if the size and the shape of the skull do not appear to have a particular correlation.

The combinatory arrangement of the neurons could result in a countless number of different combinations, each corresponding to specific and clearly differentiated mental states. Such diversity makes the occurrence of a misunderstanding by the consciousness very unlikely.

Neurons interconnect following the requirements to build evolutive cellular configurations which migrate toward a consensus — that is to say, a stable thought able to reach the consciousness. In any case, the brain's activity which we are able to detect electrically from outside constitutes only a limited share of the total electrical bustle, called disorderly even if purely deterministic.

So what exactly does "to understand" mean? Assuredly, this refers to the ultimate mechanism of consciousness, much more developed in a human brain than in an animal one (depending on what human and what animal, which indeed remains to be checked). It is the ultimate step which allows us to evaluate, to guide, and to make a choice between several possibilities; to understand also allows extrapolation into a possible future.

# Cognitive Organization

## Cognition

Thoughts wander at the boundaries of the unconscious, but they can also be controlled by voluntary attention and alertness. Thoughts can even become logical, deductive, ordered, "rational," driven by a controlled will.

The will (a concept worth being defined and materialized) is able to induce pressure on the subconscious to bring to light the elements which instant thinking requires to build a virtual but coherent reality. Such a synthetic or critical effort can even induce the establishment of a reliable value judgment or a provisional result which will be categorized in good order in the event of any need.

Consciousness also may obey irrepressible, irrational, illogical (*a priori*, but who knows?) impulses we call instinct, inspiration, illumination, or even vision because there is not enough awareness of its origin or purpose.

Consciousness is at the heart of the whole cerebral organization; it is like a central "router" of our personality, which governs our alerts, our decisions, our reflections, or our inspirations. It is the prelude to every mindful or sometimes instinctive action.

Its localization in the brain is disseminated, but a high concentration of the neurons which are responsible for consciousness has been identified especially in the prefrontal cortex. Its way of operation remains unexplained and undecoded in spite of the efforts undertaken, because of the particularly complex, changing, sometimes apparently illogical behavior. It implies an appealing multitude of links and much-diversified exchanges over unidentified paths with every area of the brain, which constitutes the domain of the unconscious, not so easily accessible and in essence still little-known. For instance, one may wonder how the unconscious works. Does it use words as consciousness does?

However, the MRI study of the visual cortex has shown a dedicated activity related to the degree of consciousness; a "flashover" similar to that produced by an image or a sound.

It is unclear as to what the exact scenario of the biochemical process which makes us think of this preferentially to that. Is there, as suggested by Descartes, some hidden demon to guide the choice or is there pure randomness; what does it mean to be free to decide and to what extent? How do ideas pop up in our mind? What is the criterion for reaching the field of consciousness? There are some questions to be answered before we try to copy a mind. Would it be a way to improve the process of getting "good" ideas?

As far as we know, consciousness occupies a rather limited place in the brain compared to memory and even to the unconscious activity which is constantly bustling (be the consciousness active or not). A permanent and essential relationship with the unconscious is steadily maintained.

All this knowledge that we have accumulated is still basic and piecemeal, and that clearly shows the complexity of the neural mechanism that the accessibility in awareness implies. However, we do outline the main directions. The aim is to check the reactions of consciousness to known stimulations in order to understand the functioning and explore the barriers between consciousness and the unconscious domain where the essential brain activity takes place under the supervision of some cortex areas specialized in the thinking operation. The smooth functioning of the operations is to be supervised carefully.

If a digital copy of the brain is to be expected someday, consciousness should be the key element to bring up and we do have to get a better understanding of it before getting down to work. Marc Bishop, a computer science professor specializing in cognitive software at the University of London, affirmatively said: "A computer cannot have a consciousness." Is that true and will that remain true in the future?

## Memory and Oblivion

Obviously, memory is the unavoidable counterpart of consciousness. The volume it occupies in the brain is assuredly larger than that of consciousness, but remains relatively small in comparison with the whole unconscious domain.

It is usual to more or less distinguish some categories of memory:

- The register of sensoriality records in a transitive way (some milliseconds) the information from the sensors of the five senses.
- The running memory is more directly concerned with the continued consciousness.
- The short term memory (short term store) plays a determining role in cognition.

But it is the long term memory which concerns us more directly in the framework of consciousness. It is the place where available information is stored.

Obviously, not all the information received by the memory is worth being preserved and consciousness has to validate the acquisition, adding a complement about the importance of memorizing it, with more or less confidence following the estimated importance of it. The will can play a decisive role in the secured memorization.

The memories are stored in the form of neurons, or a group of neurons — neurons connected through the corresponding synapses. A biochemical process controls this synaptic "plasticity," leading, with time, to the demise of these connections if they are not requested frequently enough. This is the oblivion process[9] which eliminates the unused or rarely-used connections, because of their lack of interest.

---

[9] A comparison with a computer does not hold at this level: bits can be added much larger and faster than any hippocampus can grow neurons!

The brain is the only judge of the opportunity of this cleaning operation, which generally occurs during REM (rapid eye movement) sleep. This allows us to perform a recycling of the neurons toward other uses or, more simply, achieve an energy saving.

The more the information is used, the more it is secured. The reason a memory vanishes is either that it has been effectively deleted or that the access path is off-duty. To get things back on track, one can try, in a voluntary effort, to look for nearby information (facts or suggestive contexts) lying in the memory; this could reactivate the path. Oblivion can also originate in a lot of bearings related to the acquisition of the information.

Note that this will in no case affect a computer, which never forgets anything (in principle)!

Memory is a fragile structure especially sensitive to aging and neuro-degenerative diseases such as Alzheimer's. The loss of neurons leads to a reduction in the mental functions and the memory (loss of recent memories, cognitive troubles, etc.), causing cortical atrophy and destruction of the hippocampus.

## About Animals

The consciousness is in no way an exclusive privilege of the human brain; animals too are conscious, and this is necessarily related to the requirements of life. Of course, its relationship with "intelligence" is to be considered differently.

Animals like the mouse, dolphin, or chimpanzee (which is close to humans) are widely studied in laboratories in order to learn about the organization of their neural systems. Of course, the performed experiments differ from that on humans because communication is not so easy between the observer and the observed, but the counterpart is that putting a probe in the brain of a mouse does not pose an ethical problem!

Darwin was of the opinion that animals show various forms of consciousness adapted to their being and the necessities of their survival. Neurobiologists call that "phenomenal consciousness" or "core consciousness" (Antonio Damasio).

Some animals can display mental flexibility in adapting to a new situation, in a reduced speech process. Some are also able to use elementary tools. Elephants have an especially large brain (about 4 kg) and clearly show a conscious vision of the world and, some say, an unfailing memory!

## Toward the Discovery of Consciousness

The mysteries of the neuronal mechanisms of consciousness are just beginning to clear up. However, we still are far from an overall explanation. One of the most important points is the access to consciousness, the main electrical expression of it being an especially intense P3 wave.

However, the "brain Web" is far from being decrypted, because everything is mixed in a general pandemonium (with respect to our rationality). No machine will be able (but never say never?) to describe in detail what we are thinking about right now.

### The Trial-and-Error Method

To reach the level of the consciousness, information arising from the exterior requires a short time which corresponds to the necessary delay for the unconscious to evaluate and provisionally store the elements. This makes it necessary to keep the information active and stable during this while.

If the information is not validated or has been changed in the meantime, then it is not transferred to consciousness but a record is kept for a moment in the subconscious. Then the information is said to be "subliminal," and this is an experimental way to investigate the means which the brain uses to activate the consciousness. Of course, only the consequences are revealed, and not the causality of the processes.

Experimental approaches have been developed to take advantage of these subliminal mechanisms to probe the subconscious domain, which we have no direct access to.

In any case, the consciousness is unable to take into account more than one object at the same time; there is a selection procedure which makes the choice — even though other competing ideas stay in the background (say, the preconscious field), ready to be brought to the surface. There is a permanent alertness or vigilance; a permanent instinctive procedure of trial-and-error evaluation takes place.

Also to be noted is the possibility for the consciousness to be alerted about a forgotten element which is supposed to be of some importance but cannot spontaneously be brought back to mind. We say we have it on "the tip of the tongue," but nothing comes. The trial can be helped by some intellectual gimmick as a "reminder," a word close to the subject, an object, or a knot on the handkerchief; then the consciousness is able to make intuitive correlations in the subconscious, which brings to mind the right elements and leads to the solution. If these recipes are ineffective (and this is often the case when one is aging), it is recommended not to think about it for a while and, very often, the answer pops up in the mind all of a sudden, without any effort to get it.

Consciousness is not restricted to a single specific place in the brain but is, rather, distributed throughout the brain in order to keep close to the active zones of interest. It retains for examination any element which represents a noticeable interest in correlation with the followed idea ("fame in the brain"[10]). It rationally coordinates disparate features which could be significant of something.

To reach the consciousness, several cerebral regions have to be synchronized in a global neuronal state of excitation. Then the flashover starts and is powerful enough to be easily detected from outside. This signal validates the information, which then goes beyond the domain of simple reflexes.

The cortex is constantly carrying a backlash which spontaneously self-organizes (William James: the flow of consciousness). It is this random process which produces the ideas sent to the consciousness.

The access to the consciousness comes from a sort of information which needs to have some relevance and consistency before being validated for this purpose. This process of maturation may take some time before being allowed to reach the consciousness.

This is because of the complexity of this information processing, which implies back-and-forth exchanges (loops), that the field of consciousness is very narrow: only one idea is to prevail at a time; even if several "scenes" could superimpose (an image or speeches, for instance), a single one will dominate. Attention will focus on a single idea even if the rest are the subject of alertness.

---

[10] *La conscience expliquée*, Daniel Dennett, JACOB, 1993.

Alertness is a voluntary kind of attention which raises awareness while waiting for something well-defined.

The attention, especially the selective attention, is another unconscious parameter which accompanies a claim for pertinence before validation.

There also exists a "pre-consciousness" state, which is a separate checking prior to the triggering of the consciousness itself. This is an intermediate state of the unconscious which allows checking the grammar, memorizing key sentences, calling back similar connotations or priorities, recording mnemonic rules, etc. This last point is often important for recalling something we usually forget: a pronunciation, a song, the rule of three R's, etc. Words are mixed with images or sounds. It is usual to say "I don't see what that means" or "I am looking for similitudes" or "at first glance, I should say," which are expressions that in no case refer properly to sight.

The biological preconscious logic is able to discriminate analogies, assess a metaphor, and get help from a word correlation.

The management of the information given by the five senses corresponds to particularly elaborate operations which give rise to still unclarified issues.

- Three-dimensional vision results in particular from highly elaborate information processing worthy of advanced software in a supercomputer. Images are provided by the retina (which is an integral part of the brain); between the sensorial detection and the conscious perception, an intense work of unconscious implementation is taking place. Both images apparently reach the brain in total disorder, and they have to be reconstructed from scratch with data processing that is much more refined than can be done with software like Photoshop. Each of the images must be assembled, recalibrated, and renormalized by eliminating the shape and intensity distortions induced, for instance, by the ultrasensitive fovea, the surrounding macula, and the blind spot (which has to be eliminated) in order for the two images to be put in perspective and exactly matched to give the sensation of 3D. This acrobatic processing, of course, has to be completed in quasi-real time, to boot. On top of that, the eyes automatically follow in synchrony the direction pointed by the fovea and selected by the brain in a perfectly unconscious operation.

- Sound and speech recognition also are very complex operations which require data processing which is close to what can be performed by a computer but which cannot be achieved in analogical signal processing. The words to be borne in mind are extracted from the noise with a kind of pattern recognition procedure; they are identified (with the memory), sorted, and again put in order; the sentence is then reconstructed. Only after these steps are completed and the sentence considered being consistent is the consciousness authorized to "hear," with the corresponding associated meaning.

In electronics, the analogical sound processing usually proceeds (as far as possible) from band pass filtering to separate the message from the noise component or other identified unwanted components. More cannot be achieved.

Digital processing and software allow one to dig a bit more by identifying and separating the sentences (speech recognition or reconstruction of an artificial voice). However, the brain is able to follow and separate (to some extent) a given voice amid noise or even other simultaneous different voices while being attentive to any peripheral information. The computer is not yet able to do the same "channeling" and nobody knows how the brain manages this situation so well.

## Troubles Out of Control and Disorders

However, the consciousness can be misled or duped in some circumstances, independent of the will. This is the domain of illusions.

Dreams, of course, are an ordinary situation where the consciousness is out of control except for the underlying alert of any uncommon external event. This is a relaxation state for the brain, which is elsewhere busy cleaning up and rearranging the memory, as can be done in a computer with disk defragmentation.

Hypnosis is a more complex state of lost control of the consciousness. It can be induced by an external person or even autosuggested; the exact mechanism is still unknown, but one must note that it works. It is now admitted that this state is different from sleep and is considered to be an open door to the unconscious. This strange behavior is shared with animals. I have personally seen chickens being hypnotized under the gaze of a person. Some speak of a "surconscious state" close to sleepwalking.

There is no doubt that hypnosis is deeply connected with the deep brain. It has been established that hypnosis can help to cure phobias or addictions, or induce false memories.

Would a computer have to simulate such surconscious states?

A great number of categorized disorders are also specifically related to consciousness under exotic names such as "the neglect syndrome," "apathy," "akinetic mutism," or "anosognosia." They all are related to psychiatric problems.

It is now established that residual "minimal consciousness" persists even in dramatic situations like "the locked-in syndrome" or post-traumatic unconsciousness, and that a normal state may sometimes be recovered. Re-education operates as a learning process. All of that obeys a deterministic logic, and new neurons, which are continuously generated in small amounts, can possibly also contribute to the recovery.

All of this organization results from training of the brain accumulated from birth. If anything artificially changes the normal order of things, then the consciousness is lost.

## The Experimental Means

Consciousness has become the subject of a particular science which claims to be exact. MRIf constitutes a tool for investigating the brain to the point where currently it could become a reliable "detector of consciousness" with this essential question: Would people who fall into a deep coma or a vegetative state be in some way conscious? The answer is: It is likely in many cases. This has undoubtedly been proven. Brain damage can be localized in the motor area, whereas the rest of the circuits still work normally. It is also known that the brain is sometimes able, to some extent, to reconstruct circuits which bypass the wounded zone by itself. Unfortunately, we still know very little about this process and the way to stimulate it. So what to do with palliative care? How do we decide that the point to stop intensive care has been reached?

EEG is the preferred tool for exploring the mind because it is perfectly non-invasive and very easy to use. Significant strides have been made in the detection and the processing of the brain's electrical signals; some say we are currently able to "read in the brain"!

The state of consciousness starts in one of the specialized zones of the brain (music, speech, image, smell, or even a memorized idea) and after validation gives rise to a general "flare-up" of the total cortex (frontal and parietal lobes) which can easily be detected and followed by EEG.

Another way of proceeding, which is less physical but is considered to be equally scientific, consists in making use of a panoply of controlled stimulated illusions (visual or aural) to dub the consciousness and afterward study the so-induced reactions, if any.

Subliminal tests are now well developed; an arsenal of masked or deceptive sensations can be used to explain the reactions under the control of an observation facility. Subliminal thresholds are obtained by suggesting sounds or images so briefly that they cannot reach the consciousness, whereas they are well perceived by the unconscious. This is the frontier of perception. In these neuronal recorded reactions, the most difficult part remains to establish the causality relationship between the stimulation and the triggered profusion of epiphenomena which occur all around.

Cognitive psychology demonstrated the importance of subliminal perception — this shortcut to the subconscious that consciousness lacks the time to capture and yet still leaves a trace. An extraordinary power of calculation remains at this level. Poincaré already thought that the "subliminal self" dominates the conscious one.

Obviously, the "conscious self" gets the last word, but some are of the opinion that this could be an illusion; we could be subject to our subconscious and we may be the playthings of it, which questions the relevance of our free will. Could we be unconsciously programmed? Terrible perspective!

So there exists a subjective consciousness which is able to spontaneously suggest such virtual illusions, hallucinations, and visions, and which could induce pathological artefacts such as hypnosis or even schizophrenia.

## Human Logic and Intuition

As a new science in the neural domain, cognitive psychology tries to introduce some rationality into the chaos of our cerebral behaviors. Thinking means reflection — that is to say, back-and-forth transfers between consciousness and the unconscious domain to select the appropriate elements. Who should be the judge of the opportunity of the conclusion? What exactly is the role of the will in this procedure?

This could be paraphrased using the saying "The early bird catches the worm." That is undeniably true, but don't forget that "The second mouse gets the cheese"!

## Subconscious Consciousness

The title of this subsection could appear somewhat questionable at first glance. What I mean is that there is an intermediate state of consciousness which provides us with especially rich innovations. This state occurs in the silent darkness of the morning, when one slowly emerges from deep sleep, the eyes being still closed; all of a sudden, a new bright idea you were looking for or a clear solution to a pending problem pops up in the mind, coming from nowhere and without any apparent reason. Any active-minded person should be able to confirm such an unexpected experience.

Then it is worth swiftly writing some keywords on a piece of paper in order to secure a link with the memory. If not, in spite of the effort to remember, the information will spontaneously disappear in a short while into the depths of the subconscious, as soon as the awakening process occurs; bringing this precious thought back afterward will be very difficult if not impossible — and that's a pity!

The consciousness occupies a rather limited place in the brain where the virtual ideas have to be housed also with the unavoidable management of the external world sensations.

## The Role of the Words

The word "consciousness," a long time ago, was a taboo word, due to the difficulty of defining it clearly. Today things have changed and there are now deep concerns about it in the neurosciences which aim to become experimental.

The role which words play in the consciousness is essential; we currently know that the "identity" of words originates in an area of the brain called "Broca's area," located in the left parietal lobule, but the way it works is still unclear. The unconscious governs and the corresponding neuronal code is very specific.

Consciousness uses raw data (from images, sounds, sensations, etc.) but, more important, it also uses thoughts which are described in the brain as words. They spontaneously pop up, and logically and sequentially structure our reasoning, our deductions, our interpretations, our literary creativity.

This supposes not only a dictionary of signals (words) but also the development of the meaning of each word through correlations between words, the evaluation of the semantically close-lying words (synonyms), or the detection of colloquialisms and derivatives. All of that is not so simple and makes especially difficult the elaboration of an intelligent language — that is to say, one which makes sense.

Logic has to be weighted by a host of related elements with varying levels of importance that are difficult to assess although the brain takes it easy. This natural linguistic approach can be found in some rare and specialized dictionaries (in French, "Robert"). This continues to a long education in our native language. Those who did not benefit from such an advanced learning in the use of words (illiterates) have serious mental development impairments and will never be able to perform intellectually.

The oral language structures our state of consciousness, not only to speak and hear (which could be done by a computer) but also to think and understand — that is to say, evaluate. This fundamental ability is the bases of any thought — even the unconscious. Signals similar to those used in language have been detected in anesthetized patients, or patients who had fallen into a coma.

Epigenetics is a very new science, devoted to the changes in the gene activity which can occur without changing the DNA sequence. This could play a role in the acquisition of a new language and the induced new neuron organization.

To learn a new language requires one not only to memorize the words but also to deeply re-educate the brain; you really speak a new language when you are able to directly think with the foreign words. Thinking and reflection are done with spontaneous words. The computer will "think" when it will be able to actually express through words by itself.

Language means training, as in everything; there is perhaps hardly any innate aptitude — only acquired learning guides us, learning which has accumulated since we opened our eyes for the first time. The hard drive was empty and the BIOS was limited to the survival reflexes and the largely open "learning process." We are terribly dependent on the education we got at an early age, which left an indelible mark. Our personality grows in the first few years — hence the great importance of parental care and family closeness in elaborating a "good" brain for children.

The basic reflex for a beginner in a new language is to think in his natural language and search his memory for equivalent words and then try to assemble them in a convenient grammatical order; this is mental work which is a bit like the mechanical logic of mental calculation but which is not enough for one to really master a second language. Such a complete control can only be achieved when the subconscious education is sufficient for one to directly think with the new language, as a substitution effect.

Translation is in no way changing a word with another in the new language; it is more likely the transfer of the very meaning of a full sentence into a different formulation preserving the intended meaning. As was said before, the words do not represent a self-identity but rather a volume of relationship we called the "sphere of meaning." These implied references can assuredly be quite different from one language to another even if the respective words are taken as equivalent.[11]

An extreme example of this can be found in comparing the use of the word "ignorance" and its opposite, "knowledge," in European minds versus Arabic ones. In the former case the words refer to the whole assembly of sciences and intellectual culture, whereas in the latter case they essentially make reference to only the assumed knowledge of the Koran.

To be efficient, this cerebral implementation of a new language (when successfully achieved) has to operate alongside the previous language in a flip-flop manner. This confirms the priority of the subconscious (a central executive process) in the cerebral activity. Then, in the future, if we have to copy a human brain in a computer, it will be mandatory to know in detail its hidden behavior, its own internal biochemical "cooking."

---

[11] This ambiguity can even be used in a perverse way to induce people to go in a wrong direction by using intuitively biased words.

## *Intuition*

The brain is not limited to commuting stimulations — it also creates its own internal paths. The computer, for its part, does not bother with the internal cooking of the electrons, which only follow the orders. To really simulate a brain, the computer will be required to understand by itself the meaning of what it is doing. But it does not yet know how to proceed; software consciousness does not yet exist.

There still are, of course, unreachable mysteries in the conscious approach, such as the instinct or the innovative creation, the triggering of which lies in the depth of our unconscious cerebral life, but they surely arise from an undetectable coincidence and determinism. Jacques Monod[12] said that randomness and necessity are both directing our biological destinies.

Would a machine be intended to wake up to be conscious of its thoughts, to develop a judgment? What could be its action and decision autonomy? Would consciousness be destined to play a part in the transhumanism? Pending questions!

---

[12] *Le hasard et la nécessité*, Jacques Monod, Editions du Seuil, 1970.

# Chapter 4

# Transcendence

What definition could we give for transcendence? Metaphysicists say that this word refers to anything which lies beyond the domain we take in reference, which is located above and belongs to another nature, after the perceivable and the potential feelings of human nature. In any case, transcendence is a subconscious domain which is not spontaneous but results from teaching. The most famous examples of transcendence deal with mathematics and with religion, both of which are a mental creation more or less embedded in the depths of our self.

For the phenomenologist, the transcendent is what transports our consciousness above our objective perception, beyond any possibility of knowledge (Immanuel Kant).

The newcomers, the transhumanists, refer to everything which confirms or surpasses the concept of the technological "singularity," which they say is to come soon. There will come a time when technology will be able to progress by itself in its own elaboration and its complexification; humans will adopt an attitude of confidence and surrender to the self-sustained technological progress.

Imagination and abstraction always leads to transcendence, this part of our self which is not directly related to the perception of the world out of the five senses but lies at the limit of our intelligence and consciousness. Would the humanlike computer have to address this domain and include a module of transcendence and its more dedicated extension, religion?

## The Self, The Soul

This is the limitless domain of the philosophers and the religious. "Who am I?" Followed by "I think, therefore I am." There are the famous basic questions, still unanswered. The self is often combined with the soul in Western countries. Could it be reduced to a computer concept possibly diluted in the "cloud"? Would it be vulnerable to hacking? Would it be sensitive to brainwashing? Also to be considered is the permanent evolution of the self, depending on the external world influences. Our self is not the same as it was last year, even if we are not clearly conscious of the change, and it will not be the same next month. The brain evolves constantly.

Would a soul be conceivable in a machine? Frankly, the question itself should have been considered crazy a few years ago; what about the current opinion? A cautious answer could be "Let us wait a bit; it would not be surprising in the medium term."

Transcendent things relate to conceptions lying out of the direct logic of the material world. They are embedded deeply in our subconscious but instantly re-emerge at will. What we consider as our "self" is the materialistic version of what the idealists call our soul, giving it possibly a vocation to be eternal; nobody knows how it could be there but everybody is convinced it is there, or in a connected parallel universe, maybe. The self, at most, is assuredly the favorite object of the psychoanalysts (the self or the superego of Freud), but the soul is more elusive and fundamentally it is the favorite field of the religions.

This is the very nature of our personality: fragile, versatile, evolutive, unique, this virtual structure is of our own belonging, our ultimate identity, our character.

It is known that drugs can play a role in our behavior to such an extent that our very self could be changed. Other means are currently also emphasized; they use electrical stimulations or implants which can do the same. The more we know about our brain, the more we try to "put our fingers in." The first excuse, of course, is to cure apparent disorders, but why not go further with a computer which is deemed to be competent enough? Where does that stop?

# Transcendence: Mathematics and Religions

Beyond these considerations, transcendence causes us to enter virtual worlds extrapolated from reality. Basically, in these virtual worlds of deep thoughts, we find mathematics which obeys a strict deductive logic from limited presupposed axioms, but is also concerned with more virtual fields of the deep mind, close to philosophy.

Also, of course, there are the more esoteric religions which have not any causal relationship with reality and rely only on our own convictions and education.

It could be said that transcendence results in a teaching of the subconscious to introduce indisputable "truths".

## *The Mathematics*

The story of mathematics is deeply illustrative of how a transcendent construction can be brought into the mind and takes place in the memory and the subconscious, to get ahead.

When the first glimmer of intelligence started to illuminate the thick skulls of the Neanderthals, their first existential question surely was "How?"[1] How to cleave a flint pebble, how to make fire, how to sew a garment, how to find a cave? This proceeded from the direct relationship with the rough surrounding physical world.

### *The numbers*

Then the millennia went by and this primitive man (on the verge of discovering capitalism) ended up asking the fundamentally more specific question, "How many?" And then this man made an association of ideas with … the fingers of his hands (which he always held with him, and this was very handy). This intellectual step, this pure intuition, then belonged to pure genius. It was the first symbolic transfer of human abstract (transcendent) thinking toward physical benchmark.

---

[1] The English language is remarkably subtle in making a precise difference between "how," "how many," and "how much."

Mathematics was born from this essential need to count, at the very basis of any business. Much later this primitive genial idea of the numbers will turn into the sublime intellectual quintessence of the human species we call mathematicians (!). The ancient Greek Mathematician Euclid, "the African" in his *Elements*, gave this definition of the numbers: "A number is a plurality made up of unities."

Quickly afterward a fierce competitor arose: he was the man from the Cro-Magnon, who certainly was a Briton as he appeared not to behave like everybody else, and he asked differently: "How much?" The problem was no longer to count objects but, more specifically, to evaluate a quantity — that is to say, make a measurement, which is quite different. Then it became possible to measure (i.e., compare) a jar of oil, a bushel of wheat, the length of a rope, after having previously defined a unit (and an origin). The physicist was born! He was more concerned with the real world and corresponding realities.

"History began with Sumer" (3,000 years before Christ), said the mathematician Kramer. Figures, such as counting figures, actually appeared among the Sumerians before letters. Pythagoras, for his own part, added: "Everything is numbers." However, the purely abstract notion of numbers appeared significantly later. All of that, even though obvious today, was the most difficult knowledge to elucidate when one was starting from nothing. Genius, inspiration, and logical intuition were required; a machine can hardly be imagined to have done that! Where did this inspiration come from? Was it a biological Darwinian determinism effect, a gift from God? No matter what, it has been done.

Afterward, the infinite crowd of numbers was tentatively divided into categories as they were discovered: fractional numbers, whole numbers, irrationals, decimals, negative numbers, complex numbers, and even "transcendental numbers." Leibniz found the difference between $\sqrt{2}$, which is irrational (no means to give an exact numerical value of it), and $\pi$ or $e$, which are transcendental (not a solution to polynomial equations).

Some ideas were difficult to accept, such as fractional numbers, which were for a long time inconceivable by the Greek thinkers. A fraction in their minds could only represent a proportion, a ratio, but in no case at all a number; they were deeply convinced that the space was necessarily empty between the whole numbers (as with the fingers of your hand).

Yet they discovered that the square root of 2 was irrational (which means immeasurable), and that caused overwhelming intellectual dilemmas and endless discussions. A real cactus!

To make things worse, the physicist invented a number still unexpected which can hardly have been imagined from scratch and which stays well beyond any reference in the real world: a number essential in the process of measurement, a number aimed at representing ... nothing! *Zéro est arrivé*!

The zero is a purely transcendental invention. It can be indefinitely added to itself or to any other number without changing anything, but by simply being put after a number it contributes to the decimal system.

## The algebra

A step further, it appeared that these numbers of any kind had to be related with logical links and considered as virtual entities. This discovery was made by some Indian scholars such as Aryabhata or Brahmagupta (in the 6th to 7th centuries), and it later gave birth to an extensive literature (but without any further conclusion or discovery) in the Arab–Islamic world. A name was given to this new science: algebra (from the Arabic phrase "*al djabr*," which means "to tidy up").

The very start of algebraic science was given by the Italian, British, and French schools all along the 13th century: Cardano, Fibonacci, Tartaglia, Grosseteste, Bradwardine, de Maricourt, Chuquet, and many others. But the current esoteric formalism of algebra had to wait until the 16th century for François Viète.

An example of the famous "equation" proposed by the renowned Khwarizmi and the modern symbolic version we are using today is shown in Figure 4.1. This new formulation was imagined by Viète, the genuine and genial discoverer of modern algebra. This scholar proposed for the first time the original idea of this symbolic, sophisticated but very sparse language; for the first time also, he identified the different roles of the parameters, the variables, the unknowns, and the way to write a symbolic relation we actually call an equation, and occasionally also the way to solve it (when possible). Viète was a visionary (I would say transcendental) mathematician but, unfortunately, he is not celebrated today with the praise his work deserves.

فأما الأموال والجذور التي تعدل العدد فثل قولك

مال وعشرة أجذاره يعدل تسعة وثلاثين درهما ومعناه أى مال اذا زدت عليه مثل
عشرة أجذاره بلغ ذلك كله تسعة وثلاثين . فإمه (٠) أن تصف الأجذار وهي في
هذه المسئلة خمسة فتضربها فى مثلها فتكون خمسة وعشرين فتزيدها على التسعة
والثلاثين فتكون أربعة وستين فتأخذ جذرها وهو ثمانية فتنقص منه نصف
الأجذار هو خمسة فيبقى ثلاثة وهو جذر المال الذى تريد والمال تسعة .

The famous "equation" of Kharizmi

$$x^2 + 10x = 39$$

The same after François Viète

**Figure 4.1.**   The birth of modern algebra.

With all that basic stuff, mathematics took centuries of transcendence to reach the sophistication that has currently been achieved.

The world was not made in a day ... nor was mathematics. All that intellectual construction has taken place in the human mind and could possibly be transcribed in a computer — but what about teaching the machine how to proceed further? There are no recipes available for nurturing imagination.

However, mathematics (from Aristoteles to today) cultivates a closer relationship with philosophy through the concept of "logic". In the opinion of Kant, "logic should be considered as the science of judgment," and this is valid for the philosopher as well as the mathematician.

Symbolic logic is the study of the symbolic abstractions that capture the formal features of logical inferences. The form of an argument is displayed by a sentence in a formal language which can take very diversified and complex shapes (grammar, ambiguities, etc.) to the point where it becomes hardly practicable to put it in an inference.

Recently, Hilary Putnam has raised the following question: "Is logic empirical?" Things sometimes have to be reconsidered from a more transcendental point of view in respect of what is traditionally taken as evident; famous examples of that can be found for instance in the concepts of "relativity" or, more generally, in quantum mechanics.

In the same vein, Gottlob Frege and Bertrand Russell (philosophers), who advocated "logicism" claimed that mathematical theories were purely sophist tautologies. This did not prevent "intuitionistic logic" from being used in computer science. All of that stays in a very transcendental controversy.

## *The Early Ages of the Religions*

From time immemorial, humans, as they are now, were very unequally intelligent; some of them, a bit more shrewd, intuitively understood that, in order for a group to be and stay coherent, it was necessary to keep the minds engaged in the same direction by the means of shared and indisputable convictions or myths.

The need soon appeared to create a model of basic beliefs generating homogeneity in the group and a strong feeling of belonging to the same community. This was intended to provide a common reference for what is good and what is bad — a moral code of good social conduct accepted by everybody. This model had to infuse the mentalities in order to channel the behaviors toward reflex mindsets which could help ease sufferings and propose simple mental solutions to the difficulties of the rough temporal life of the time.

In this aim and in order for the message to be well accepted and easily memorized, the shamans, prophets, and other kinds of druids all referred to the surreal, the legendary, the imaginary, the mystical, of an unreachable world. In some way this was a kind of soft brainwashing which was intended to benefit the whole community. Of course, this transcendental construction remains, by nature, inaccessible to any objective observation. The times were to legends, tales, and other phantasmagorias so pleasant for a romantic and wandering spirit. Mythologies were born under various aspects and were readily accepted. There were no other ways to entertain people (TV had yet to be invented).

With these "teachings" a subtle equilibrium is managed between the promises of a bright future (after death, however) if we are behaving well (that is to say, following the established prescriptions) and the assurance of a dire penalty in case of disobedience. Divine law always adds to the social conventions to keep a balance; on one hand, the spiritual and on the other the secular, each within its own area, but with the prevailing religious recourse.

To be sure that the message was well printed in the depth of the mind as an uncontestable truth, the children were a primary concern of this teaching. Globally the recipe worked satisfactorily, and it was because of this psychological mold that civilizations were able to develop coherently. All of them were based on religious unity, in every country, whatever their shape, from Mesopotamia to Central America or Asia.

Religious requirements were highly diversified in their intensity, from the peaceful Buddhism to the precise Judaism, the rigorous Christianity to the intolerant Islam — this, assuredly, depending on the very character of the concerned populations. The limited mobility of the populations in the old times certainly contributed to the peaceful cohabitation of the peoples, apart from some violent episodes, such as the Arab conquests or the arrival of the Spanish Conquistadors in Central America (who likely unconsciously imitated the previous Arab invaders, *mutatis mutandis*).

The recipes worked well and improved over time. Some versions were so nicely designed that they endured and dominated while improving. Obviously these convictions, generalized in the groups, found themselves in competition with those of neighboring groups; internal divisions or merging, even wars were needed when no other accommodation was possible. So the dominant religions on the world scale were established.

Of course, all of this is based on convictions that have to be accepted as a *sine qua non*. A critical mind is intrinsically incompatible with any religious belief. Examples of opposition are numerous and famous (Galileo, Darwin, etc.), and they followed step by step the evolution of the sciences.

Through the highs and lows, this worked well over the centuries when the way of life did not appreciably change in a human lifespan. The means to know more about our origin and future did not exist until recent times, and this is what has abruptly and actually changed presently.

## Nowadays, Modernity

Everything changed with the coming of modern societies. The sciences never ceased to batter against all of what the religions presented as indisputable. This has succeeded in generating today some fundamental revolutions in the more intimate convictions we are born with. The material and intellectual possibilities opened up by science and technology make us dizzy; there is no longer a religion to cling to. We have entered into an unprecedented knowledge of our universe, from the remote limits of the sky to the tiniest limits of the infinitely small.

This has been made possible due to increasingly effective instrumentation and computers which allow revelation of the secrets of the universe. There is compelling evidence of the discoveries which upset the previous, unfounded beliefs related to religions.

Biology is especially evolutive with the studies of the cellular behaviors, the discoveries of genomics, the life processes. This currently poses moral questions which religions are no longer able to answer: *in vitro* (even purely artificial) procreation, genetic mutations, extended life, biologically modified humans, artificial intelligence, etc.

Also, an emerging problem is the possibility for humans, in a not-so-remote future, to be in direct competition with a higher artificial intelligence. Would robots be taught a religion, if they are to be exactly similar to humans? Or would humans have to be freed from any religious links? At the moment this religious support is still essential for most people, even if its influence is slowly vanishing and even if some claim to be secular and do not officially trust in a god.

The virtual world is on the verge of entering our daily lives, to stir our minds; would our real world of the living be intended to become purely transcendent?

Conversely, a large number of humans have not been capable of understanding modernity and have returned to the primitive and compulsive messages of a religion, without realizing that we are irreversibly in the 21st century; this full disconnection has led to bloody conflicts, still ongoing.

Religion develops a terrific power of conviction, sometimes leading to a real psychological straitjacket over the self of weak spirits. This results in irrepressible, uncontrollable, obsessional collective excesses and indoctrination, as can be observed with Christian-like sects or, more likely, Islamic schisms. Would a computer be capable of such drifts?

To come back to the computer, one can remark that it apparently has nothing to do with the spiritual management of a group, whether of humans or of machines. It would then not be essential to keep it equipped with a "religious module." This is to prevent a situation where the computer would have to take care of a crowd of devoted human slaves all eagerly looking for religious comfort.

## *Neurotheology — The Recess of the Divine?*

This observed recess of the divine in our personal, mental life is clearly observable in "developed countries" which are arguably more modernized. This splitting presently contributes to a conflicting situation with indoctrinated, brainwashed people desperately holding on to their fundamental convictions, to the point of not being afraid of committing suicide.

At the same time, neurophysiologists are exploring the brain to discover how the idea of religion arises, settles, and is stimulated through the neuronal circuits. This very particular and recent research field is called neurotheology. It has already been discovered that the idea of religion is clearly related to the subconscious, where the normal situation of consciousness is damped by an external, favorable environment. It has been known for ages that religious practices can turn into psychologically disturbing situations but the innovation is that, currently, the perturbation can be detected and followed in the brain by precise, dedicated measurements.

It has been established, for instance, that the environment in a cathedral where the sounds of organs are reverberating favors mystical stimulation of religiosity, independent of any will, and triggers a cerebral activity of meditation and contemplation that closely resembles anxiety. In the old days, the walls and windows of cathedrals or even shrines were richly decorated to help create a unique, inspiring setting in an atmosphere of incense, but nowadays new churches resemble cafeterias![2] This confirms in some way the recess of the divine, which is not yet stimulated.

This religiosity has been demonstrated to play a direct role in depression or anxiety and is related to the "plasticity" of the brain.[3] Mystical ecstasy can be detected by MRI — and a specific "God's gene" would

---

[2] Hopefully some invented the Gospel song to keep in touch with God!

[3] *Introduction à la neurothéologie*, Camille François. Available at: http://www.lesmutants.com/neurotheologie.htm

have even been discovered![4] All of this confirms the new, technical approach of religious feelings.

Strong disapproval of these researches is to be noted in the Catholic communities, but Buddhists, such as the Dalai Lama, readily accept these innovations of "cognitics." Rationality, until Pythagoras, as already discussed, hardly entered into the religious framework. How is the computer to handle the Bible or the Torah? Google ignores God even though it says "Don't be evil" to its researchers!

As Pope Benedict XVI put it,[5] the present abundance of virtual images leads to: "... become independent of reality; it can give life to a virtual world, with several consequences, the first of which is the risk of indifference to truth." As a matter of fact, he does know what's what because the Catholic Church largely used the images for centuries to "proselytize" in the transcendence. Competition has now become severe!

# Modeling a Brain

All those elements triggered, in some distinguished spirits of our time, the opinion that man will go under and that robots will eventually dominate.

## Growing a Brain

At the very early stages of a baby, the brain and the body, as in a computer, are ruled by a minimum "BIOS"[6] which is devoted to elementary functions such as feeding, breathing, and moving. Neurons rush into the brain at the beginning of the embryonic life, and then the rate begins to slow down after birth. This is a period when genetics prevails in the implementation of the brain but gradually makes room for the influence of the environment — that is to say, "self-experience."

The first evolution of the baby comes with the acquisition of words and their meanings in the language. That is the reason why this period is

---

[4]L. Alexandre: www.lemonde.fr/sciences/article/2013/11/11/dieu-adn-et-depression; also http://rue89.nouvelobs.com/2007/05/24/le-gene-de-dieu-revelation-ou-heresie

[5]After R.U. Sirius.

[6]Basic input–output system — the minimal program that starts the computer when the power is on.

essential for setting up the personality of the future child. This is a period of frenzied acquisition and organization of the mind; it is the right time to get familiar with foreign languages. Also, a boy who has not been the subject of attentive care from his family environment (especially the mother) will never recover later. The way things are currently done in our modern social environment is especially poor from this point of view.

Then, with the coordination of the body comes the imprint of the physical and affective environment, which begins to model the deep organization of the brain; genetics is no longer the driving element. One currently knows that this basic education extends until age 25 in order to achieve a complete brain and a corresponding full personality; it takes a long time to make a brain! Then the neuron production slows down and virtually stops; the basic brain is attained and may evolve, but more slowly.

A similar procedure, such as the natural growth of a brain, has been proposed to make a brain-equivalent computer at the end of an equivalent self-learning process. The starting organization implies a BIOS-like software program, and also a dedicated learning software program. Basically, such an organization should be able to evolve an autonomous intelligence on its own.

Obviously, this is not so easy to do because this way of doing mandatorily implies that the computer will be provided with a "body" and a virtual surrounding strictly equivalent to the real world, including the inputs and outputs and their counterpart: pain — a signal when things go wrong. The only difference in this situation may be that, with the computer, the timescale could be drastically reduced, and also that several "models" could be grown simultaneously or copied afterward.

Research about the direct building of the brain from scratch is also being extensively performed. But previously a precise knowledge for how the mechanisms for forming and managing ideas are established, how a brain is conditioned, and how it could be modified (and improved, of course!) was required.

The human genome is certainly not the whole story, and the whole story is not explained. Everything results from learning, and thinking also results from acquired preformatting at the unconscious level. Then, would thinking be deterministic? In some way, it could be. But genomic science is still in its infancy — maybe it could later reveal a surprising reality? To what extent are we programmed or free?

Craig Venter, who obtained the first sequencing of the DNA molecule, is of the opinion that everything in a brain depends on the genes, which would govern a kind of predeterminism. Obviously, the genes control the coding of protein production,[7] and thus they are at the center of everything in the human body. Almost every cell contains the same genes, and only part of them are active. In brain cells (producing the neurons), a third of them can be found active (to our present knowledge), which is the highest proportion in the body. However, things are not so clear and the new science of epigenetics is beginning to clarify the way the genes could be influenced by the environment and the events of life. This is in no way transcendental.

## Is Emulating a Brain Thinkable?

This idea was initially proposed by Isaac Asimov in 1950 as belonging purely to a fiction science romance, but there are now plenty of followers who are convinced that it is not so foolish. The most convinced advocate of this forecasting is certainly Raymond Kurzweil,[8] we will deal with him more extensively in the following.

Many projects on whole brain emulation (WBE) have recently blossomed, accompanied by important funding, which actually makes a "digital you" more realistic in the medium future. The main difficulty will always remain: What would we have to emulate? If we don't understand what we are trying to simulate (the world of neurons, neurotransmitters, and so on), we cannot hope to make it work in a computer.

However, some "transcendent" questions should also be answered about a possible digital "you": if such a thing is possible, would it automatically have to be updated in real time; would the copy survive the death of the "meat you" (as Sirius said)? Would this transfer require the death of the meat you? Would, then, the copy evolve by it(him)self as you would? Would the digital you be considered mentally healthy and to what

---

[7] NIH publication No. 01-3440a (2014).

[8] Kurzweil, *Singularity*.

extent? Would the digital you be conscious? Would it have a specific "identity"?[9] Would this machine be considered to have a soul? Would it be a solution for humans to escape their immanent destiny and become in some way immortals?

The questions are endless and somewhat terrifying in some aspects, not counting the implications of this "new you" in the framework of the society: Would it be legally responsible for it(him)self, would it have rights, and so on in this fictive adventure? Maybe, as suggested by Hans Moravec, this new "creature" should be considered as "mind children" of the original?

Several options were emphasized to bring to reality a "mind uploading." The first solution could be to physically "scan" the brain and copy every link in order to make a faithful digital replica of the wiring. Then grow a clone or build a humanoid robot with an identical structure. This solution of a "connectome" is studied in Oxford by slicing the brain with a super cryo-microtome and marking the neurons one by one. Of course, the "copied man" has to be dead before being "sliced"; then there will remain no means to compare the functioning of the copy with that of the original brain, for which we can know how it is constituted but not how it worked.

A second way could be to build on the development of computational neurosciences and to model the corresponding structures. Not to mention the prospect of taking action directly on the neurons in the brain.

Copy or not copy, it must be recalled again that planes were not developed by imitating the flapping wings of birds.

## What About the Computer?

Then the crazy idea came of simulating a brain in a computer and due to the fantastic progress of these machines, someone imagined this could become reality in the near future. This was considered as romantic science fiction until recently, but now even hardline scientists are beginning to speak freely about it. So many unexpected events have arisen in our time that nobody dares to definitively say: "It's fiction" any longer.

---

[9] A motion was recently proposed to the European parliament, in order that intelligent robots should be considered as being "electronic persons".

## *Kurzweil, the Transcendent Man*

The champion of this "transhumanist" prospect is without any doubt Ray Kurzweil, a former student of MIT and currently a specialist of everything (electronics, computers, AI, transhumanism, tech development ... and business). He is the cheerleader, or the guru, of "Singularians" (the brotherhood of people convinced that technological progress will lead the world to a rupture they call the "Singularity").

One cannot consider writing a book like this without devoting some lines to Kurzweil, a prophet who aims to "wake up the universe" — this in spite of the violent controversies often raised against the man as well as against his convictions and his dystopian issues.

To be mentioned: he is a man who refuses to accept the inevitability of biological death and is deeply convinced that, in the near future, life expectancy could be extended to 150 years (or more, due to the ongoing explosive progress of the technologies). This idea of life extension is not stupid at all[10]: let's imagine a baby born in our age; if God spares him, he will be a centenarian...in a century. What would have become of the state of medicine (keeping, at the least, a constant rate of progress) in a century?

Transcendence is at the origin of the philosophy of Kurzweil, who certainly seems in no way a crank. The Singularity has become a sort of philosophy of the future, still finding its way, but this did not prevent the creation of a privately funded, dedicated university in California: the Singularity University, where students are trained to "modelize" the future.

Kurzweil's book *The Singularity Is Here* was a global success, a real best-seller. Also, a film was made (directed by A. Waller), which included some elements of fiction: *The Singularity Is Near: A True Story About the Future*.

With the same visionary look, he produced another film in 2009 with Barry Ptolemy, *Transcendent Man*, explaining how he emphasized "bringing back" his father (who had died several years before) by creating an "avatar" of him. This film is not to be confused with a more recent one (Wally Pfister — pure science fiction) which appeared in 2014, this is entitled *Transcendence* but has nothing to do with Kurzweil.

---

[10] *La mort de la mort*, Laurent Alexandre, JC Lattès, 2011.

Then Kurzweil convinced Larry Page, Google's CEO, to hire him (2012) as the Director of Engineering in charge of "new projects involving machine learning and language processing." You know that the watchword Google gave to its researchers is "Get the moon!"; no doubt they will succeed with Kurzweil!

The Singularity, as presented by Kurzweil, appears as a point where man and machine are to merge in new concepts and new beings. Biology is to be reprogrammed, mortality will be transcended, aging will be mastered with new nanotechnologies, red cells will be replaced by more effective nanorobots, the brain will be interfaced with computers, and AI will make superhuman intelligence possible, including the possibility of making a backup of the mind (in case of what?). That is Kurzweil's future world.

This "Transhumanism"[11] closely resembles a religion, and Kurzweil states: "If I was asked if God exists, I would say 'not yet,' adding 'God likes powers.' This last point is a key one. People currently are afraid of a possible domination of an 'improved humanity' over the usual one, leading to enslavement of those who would not benefit from the improvement of their intelligence. They usually say: Would the first adopters of brain-enhancing technology be transcendental above the less fortunate?"

However, it must not be forgotten that intelligence does not mean "power to do." You have surely already observed that superiorly intelligent people are currently scarcely "in power to do" and, conversely, those who are scarcely are superiorly intelligent!

Nick Bostrom,[12] a new recent guru, is also worth a comment; he was at the origin of the mobilization of Stephen Hawking (astrophysics), Bill Gates (Microsoft), Frank Wilczek (Nobel Prize winner in physics), and Elon Musk[13] (Tesla Motors) against AI and the development of intelligent robots. These people are not exactly regarded as egg-headed jokers when

---

[11] *Transhumanism, A Realistic Future?*, Jean-Pierre Fillard, World Scientific, 2020.

[12] Bostrom is in the Faculty of Philosophy at Oxford University. He is the author of *Superintelligence: Paths, Dangers, Strategies*, Oxford University Press, 2014.

[13] Musk is at the origin of the Institute for the Future of Life, which aims at controlling or even stopping the research on AI and intelligent robots.

they claim: "Will we be the gods? Will we be the family of pets? Or will we be ants that get stepped on?"

Bostrom claims to be a physicist, a "computational neuroscientist," a logic mathematician and, of course, a philosopher. That's a lot! I suppose that, as a philosopher, he must be classified as an extreme sophist with an effective rhetoric.

He is of the opinion that a digital copy of the 100 billion neurons and their respective connections is readily achievable even if the acquired knowledge of the brain is not yet perfect; it is just a matter of technology, he says.

## Growing AI?

Like intelligence growing in a baby's brain, artificial intelligence is growing day by day, independently, in plenty of laboratories around the world. We will deal more extensively with the subject later, but let us already say that it takes a multitude of different paths to progress.

The AI field is roughly divided into two domains:

- Humanlike thinking machines which truly attempt to copy or at least follow the human brain's organization. The direct consciousness could be more and more carefully mimicked in the details and give rise to amazing androids; however, fortunately, the access to the "transcendent" part of the mind still resists any investigation (and this is nice). Would a machine be able to become "shrewd"? I wonder if explaining this ultimate mystery would, someday, be reachable and what the consequences of deep knowledge of our self could be.

  What would be still more worrying should be that an artificial brain could be invented based on pure mathematical models, giving rise to purely artificial thoughts, including some "transcendence," resulting from rules that are different from that governing human ones.

- Artificial general intelligence (AGI), which more often refers to expert systems devoted to specifically restricted technical issues. This does not require any transcendence — only specified data processing and corresponding dedicated circuits. There is no limit to the performances of AGI in the many domains involved; the human brain has already been substantially overshot and AGI will certainly still progress a lot.

There is an intermediate area which relies on intelligent brain implants, which already exist but may be greatly expanded in spite of the many drawbacks, such as infections, energy supply, etc. We expect, in a more or less remote future, to replace or stimulate (to some extent) parts of the brain with such circuits, with healing prostheses (memory, hybrid thought, etc.) or better-performing processors. Elon Musk initiated the "start-up" Neuralink to develop polymer wires to make minimally invasive implants installed in a brain by a dedicated robot. A massive budget was devoted to the challenge and a preliminary experiment is on the way with monkeys.

The essential issue is to manage a free reciprocal communication of the brain with an external digital machine. EEG electrodes have already demonstrated that they could be reliable and open to improvements as far as the only way, "brain to machine" is concerned.

Let us imagine a day when a brain can be fully coupled to a computer in such way that the machine can be directly driven by the mind, by the way of extending its memory and providing direct access to the Web to get information. In such a case the computer would not have to copy the brain but only to complement it, to boost its possibilities in a kind of "shared artificial brain." Why can't this give rise to an artificial "assisted" transcendence?

Of course, such an issue is already partly achieved with the increasingly sophisticated "thought command" (prostheses, exoskeleton, or even fighter jet piloting); however, the most uneasy but essential part of the issue is still to be fulfilled: the way back. Up to now, no method has appeared to "make the machine directly speak to the brain." We still have to use a middle path: voice or sight.

We are still compelled to employ the usual, softer means that Nature generously gave us from the beginning: vision and hearing (not to mention other senses), which use conventional words to communicate. Would that be acceptable any longer? It is such a lengthy and digitally incompatible a solution. Could a digital illusionist or hypnotist be elaborated to inject compatible words and images directly into the brain at the right place?

# Part 2
# Towards an Intelligent Computer

# Chapter 5

# The Origin of the Challenge

This is to announce the World Intelligence Championship: you already know of the defending world champion, Brain Sugar Robinson, on my left, so let me introduce the challenger, Battling Jo Computer, on my right. The action starts!

We are leaving the warm and soft universe of living cells to enter the cold and ruthless world of silicon. Get ready for an awesome future; these two worlds will need to find a way of living together (in peace, when possible).

## From the Transistor to Quantum Logic

### *What is the Computer Made Of?*

You may have noted that in the magazines devoted to computers nobody cares any longer about what is inside the machine, to the point where Intel feels the need to stress that it is there. Basic physics, technology, and software are so trivial that nobody is interested in that stuff except for those who do it. However, the limiting parameters still stay at this level, so let us recall some basic knowledge to get an idea of it because there are no miracles in the world.

To clearly illustrate this tremendous evolution of the technology which has exploded over the past half-century, I show you in Figure 5.1:

(a) A 5-megabyte IBM memory laboriously installed in 1956, in comparison with
(b) A new high capacity 2 Terabyte (!) USB key (real size).

(a)                                          (b)

**Figure 5.1.** Examples of how technology has evolved.

The tale started in the late 1940s, when the transistor effect was discovered but, to cut the story short, the real development occurred with the MOSFET transistors[1] and the integrated circuits in the mid-1970s which made it possible to "paint" individual transistors on a silicon wafer, at the microscale. In the same operation, millions (even billions, currently) of transistors were so assembled in a square microcircuit, around some millimeters wide. On top of that, thousands of such circuits could be copied on the same silicon wafer (30 cm in diameter) and in the same operation, thus drastically shrinking the cost of fabrication of the individual component.

The years went by and the size of the transistors steadily shrank, following the famous Moore's law, which has never failed up to now. Every 18 months, the density of the transistors doubles (without noticeably affecting the cost) and consequently so does the calculation power of the circuits. Naysayers foresee that, sooner or later, a limit will necessarily be reached, but nothing has happened so far. Each time, serendipity has allowed the necessary improvements to be made. The piling-up of the transistors in a three-dimensional configuration was explored, and implemented in some situations.

Of course, the weight, the individual energy consumption and the dissipated heat also diminish in the same proportion as the response time of the circuit, because of the more limited distances to cross.

---

[1] Metal Oxide Semiconductor Field Effect Transistor.

Nevertheless, the precision currently reached for the "painting" lies in the range of 10 nanometers,[2] close to the quantum limit of a single electron. What is the trick to go further? No doubt new ideas will blossom.

The main application of these throngs of transistors is devoted to digital processing. This means that the transistors play the simple role of an "electrical tap" which is provided with two exclusive states: on or off (1 or 0); nothing else. They are especially well adapted to be operated in a pulse mode. This is sketched in Figure 5.2.

Following the electrical state of the gate, the pulse applied to the source may or may not be transferred to the drain. A scheme of this device is sketched in Figure 5.3: the switch is on or off, and nothing else; that is the reason it is called a "gate."

These basic functions are then assembled to constitute either a fixed architecture or a software-programmable provisional one. These Integrated Circuits (ICs) can be used in a memory configuration or a microprocessor structure. The astounding progress of the ICs and software has come from this technology, which is complex in the arrangement but simple in the elementary functioning.

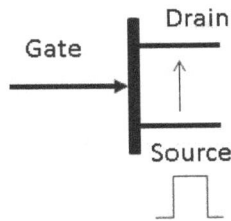

**Figure 5.2.**

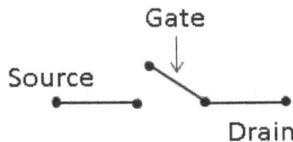

**Figure 5.3.**

---

[2] IBM recently announced reaching a 7 nm resolution on production lines.

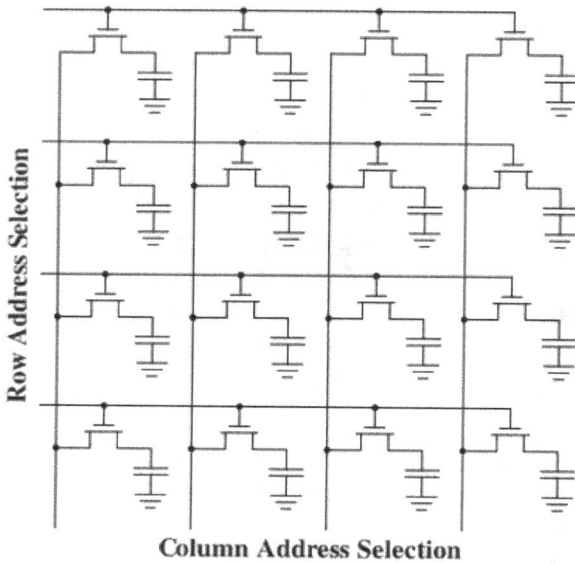

**Column Address Selection**

Simplified view of a 4 × 4 matrix of memory.

a and b

**Figure 5.4.**

In short, these individual "components" can be arranged in a matrix type organization with which it becomes possible to address any of them through a serial connection by links (called "busses"), as sketched in Figures 5.4 and 5.5.

This is the way memories are organized to store permanently (ROM[3]) or in a "volatile" manner (RAM[4]) the numbers written in a binary language (3 ≡ 00000010 in the "byte" basis of eight digits). Conversely, it also becomes possible to "read" this memory afterward and make calculations.

---

[3] Read Only Memory.
[4] Random Access Memory.

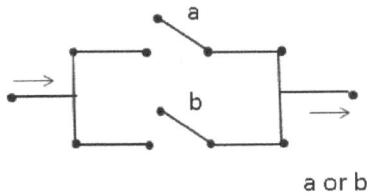

a or b

**Figure 5.5.**

| a b | 0 | 1 |
|---|---|---|
| 0 | 0 | 0 |
| 1 | 0 | 1 |

AND

| a b | 0 | 1 |
|---|---|---|
| 0 | 0 | 1 |
| 1 | 1 | 1 |

OR

**Figure 5.6.** Tables for Functions AND and OR.

This way of storing the data makes the arithmetical operations in a pro-grammable device called a "processor" (possibly "micro") very easy.

In a first step, many combinations of such simple circuits can be implemented to complete what is called "logic functions," which electri-cally translate what our brain logic uses: AND, OR, NOT, NOR, or more sophisticated others. Sketches of the functions AND and OR are repre-sented in Figures 5.4 and 5.5 as examples.

Their corresponding schematic functioning is presented in "truth tables," which summarize all of the possibilities in the input and output commands. Figure 5.6 shows the tables for the functions AND and OR.

This gives rise to the binary logic animated by Boole's algebra. Then such unitary circuits symbolically schematized are assembled to complete a full logical function and process the data. This will lead to an increas-ingly complex, but logical and purely deterministic, circuitry.

Of course, you will never closely see such a cooking, which is embedded in silicon chips and plastic wrappings with golden electrodes outside.

## Binary Logic and Thereafter

Computer logic works, like a player-piano, with signals that represent 0 or 1. There is no fantasy here; all is purely causal. The association of such circuits truly obeys an algebraic, symbolic organization.

At the very beginning of computers, before ICs were available, the technology of the time used batches of "punched cards" (IBM) very similar to the perforated roller paper of the mechanical piano or the telex transmitter to constitute a "program," whereas memories were made of matrices of small magnetic rings. All of that was rather cumbersome and limited. The initial use was to make large numerical calculations dealing with numbers translated in a binary basis automatic.

Then the process was extended to mathematical relations through the use of logical variables and Boole's algebra, which creates the relation between the mathematical formulation and the electrical binary signals. This is a perfectly "dual" operation.

Without entering into the details, it could be said that the elementary "gates" can be associated in standard, more complex assemblies which respond to more complex logic functions. This paraphernalia allows responding to the requirements of any logical issue.

Then programming a computer at a basic level consists in assembling these circuits in a convenient way, following a binary language (machine code). This is the first level of programming — of course, rather cumbersome and tedious as soon as the problem becomes large in some way. So, an interface was elaborated to automatically translate, in this binary organization, the problem previously expressed in a conventional, more practical language which was called the "assembly language."

This "second layer programming" was still not causal at all and had to be reserved for confirmed "geeks"; then, more friendly, translation was "compiled" with a third layer of Pascal, C++, Fortran, Basic, Lisp, HTML, PHP, or other, more evolved and specialized arrangements. But, to conclude the story, this was still not enough; currently Windows superposes a very friendly last layer which does not require any programming skill at all and brings the computer within the reach of everybody to carry out specified tasks of "Office" automation. All these successive virtual layers of translation have to be crossed back and forth (automatically and quickly) to get into the circuits. Every kind of the usual requirement of automation tools is provided to make the computer a close friend in any circumstance (so they say!).

All of this remains purely deterministic; no fantasy is allowed. The computer accurately obeys the orders whatever the complexity or the size of the calculations or the presentation. This is the reason why, as you may have observed, "the computer is always right"!

This is a way in which a brain works when doing "deterministic" tasks such as calculations or mathematical demonstration. However, the brain also does many things which are in no way deterministic, such as bringing new ideas, pondering, evaluating, or making reasonable choices. How to make a machine more subtle, less mechanical?

### Fuzzy Logic to Smoothen Decisions

Such a rigid deterministic behavior of the computer does not fit with the usual problems that are encountered in a computer's life. Accommodations to the classical logic of $1 + 1 = 2$ need to be implemented to simulate human decision-making in an inaccurate or uncertain context.

This is the domain of fuzzy logic, where the concept of "partial truth" has to be handled; the truth value may range between completely true and completely false, the degree of it being managed by specific functions. Fuzzy logic was extrapolated from the fuzzy subset theory, also called the "theory of the possibilities," by Lofti A. Zadeh[5] (University of California, Berkeley) in the 1990s.

As a matter of fact, before the idea of fuzzy logic emerged, Vassiliev (1910) similarly proposed considering a "multivalued" logic, for instance "true/false/possible." This was an early signal of evolution.

---

[5] *Fuzzy Sets and Applications: Selected Papers by* Lofti A. Zadeh, Richard M. Tong, Hung T. Nguyen, Ronald R. Yager, and S. Ovchinikov, eds., Wiley, 1987.

Instead of dealing with a yes or no mechanism, Zadeh[6] more recently proposed defining a partial belonging to a class of events in order to manage a gradual shift from one situation to another. This is the theory of the possibilities (not to be confused with the probabilities, even though the two are related). This should allow the management of non-probabilizable situations. To make a distinction between probabilities and fuzzy logic, let us say that probability says "how probable a variable is" whereas fuzzy logic says "how much a variable is in a set of possibilities." Purists should understand the nuance — it is obviously not the same thing!

For instance, the temperature reached by a brake could be classified as "cold," "warm," or "hot"; these three functions map the temperature scale, the incertitude being at the limits, which are not abrupt. In these regions we introduce a "degree of belonging" and a subjective distribution of probability which can be obtained, *a priori*, from the knowledge of the experts in the field (synthesis of the opinions, coordination of the objectives, etc.) or from any other reliable source.

This is close to the Bayesian approach of the conditional probability making optimal use of the available information. The starting point is *a priori* a probability of each possible world: the simpler, the more likely (Kolmogorov complexity). "The simpler" also corresponds to the shorter program.[7] The background knowledge is then incorporated step by step. New data are included little by little, thus leading to improvement of the conditional probability.

Of course, this is still quite a "fuzzy" and uncertain process — but still more realistic than an abrupt limit! The robustness of the model can then be checked by simulating tests introducing known perturbations, until one reaches an empirical confidence in the robustness of the program. This smoother procedure is closer, if not identical, to human reasoning.

In the 1970s, fuzzy commands were already introduced into the industrial processes of decision support, in the fields of economics, medicine, and industrial design.

Direct application can also be found, for instance, in recognition of handwriting or optical character recognition (OCR) of photocopied

---

[6] *Fuzzy Sets, Fuzzy Logic, Fuzzy Systems, Selected Papers by Lofti A. Zadeh*, George J. Klir and Bo Yuan, eds., Word Scientific, 1996.

[7] A program could be defined as a set of axioms and rules.

documents. A human brain knows how to process[8] uncertain (lack of data, subjective criteria, etc.) or complex, vague, or poorly defined data. To get a handle on these problems, resorting to probabilities is necessary.

This way of doing things is also very convenient for managing a self-driving car (or even a plane, as has been recently demonstrated). Moreover, this is very convenient in the database organization for artificial intelligence, expert consulting, learning, modeling, etc.

The Japanese[9] were the pioneers of the practical applications of fuzzy logic but many theoretical, mathematical developments were also proposed which made fuzzy logic a real field of advanced theoretical research. Instead of the usual numerical variables, fuzzy logic often involves "linguistic" variables that can be adjusted.

Many propositional fuzzy logics can be found as pure theoretical models (Lukasiewicz, Gödel, etc.). This, for instance, involves the notion of a "decidable subset" to be extended to a fuzzy subset.

## Formal Logic

However, the brain is still more subtle and makes tremendous use of abstract mathematics expressed in symbolic formulae. How do we bridge the gap with the logic (even fuzzy) of computer language?

Formal methods are a particular kind of mathematics applied to computer science fundamentals. The field is rather wide, from logic calculation to program semantics, and is especially applied to the safety of security-critical systems.

The Microsoft Research Inria Joint Center is particularly active in developing new tools for mathematicians, for instance to facilitate the verification of hard mathematical proofs or provide tests assessing the correctness of programs.

More simply, this allows the computer to not only give a numerical solution to an equation (as Excel can do) but, more likely, to handle the concept of variables to perform the calculations and get its formal

---

[8] Even with fuzzy logic this character recognition reaches its limits which the human brain can overtake without any difficulty (CAPTCHA characters are made to deceive OCR).

[9] The first application was on the bullet train in Sendai; see *Fuzzy Thinking: The New Science of Fuzzy Logic*, Bart Kosko, Hyperion, 1994.

solution. Matlab is a well-known and powerful software program for taking charge of the classical mathematics and allowing access to symbolic computing capability; also, Microsoft Mathematics 4.0 is currently able to solve more difficult issues.

But more formally oriented software programs, such as Maple, allow one to tackle polynomials, functions, and series, perform differentiations and integrations, and solve linear polynomial systems or differential equations in their symbolic formulation.

To go further in the competition "brain vs computer" in mathematics, we must look to Shalosh B. Ekhad (yes, this is the name of the supercomputer), used by Doron Zeilberger (Dell NJ), who says that mathematicians are afraid of computers because "for good reasons people feel they will be out of business"! And it is a fact that mathematicians have not universally embraced computers (but many of them are on the way to retirement!).

Of course, deducing new truths in mathematics always requires intuition and acts of genius to get the "good idea."

Computers are now used to make new conjectures and also rigorously prove them, even though Constantin Teleman (University of California) says that "pure mathematics does not just about know the answer; it is to understand it"; but Zeilberger adds that computer logic will "far outstrip our conceptual understanding." The day is then to come when doing mathematics without a computer will be quite impossible because of the increasing complexity of mathematics, out of the reach of the human brain. This is already especially true when one is dealing with large volumes of equations or data.

However, the main, inescapable issue persists: How much can we trust the computer regarding the search for the ultimate truth? How do we prove the proof? If the brain becomes so impotent that the human cannot check by hand the result provided by the computer, what is to be decided? Would we let the computer be the judge? The only solution would be to carefully verify the algorithm,[10] track down the bugs, check the steps, and then maybe ... rely on God.

---

[10] It was recently suggested by V. Voevodsky *et al.* that it could be possible, with their program named Coq, to formally verify the algorithms, step by step. This development could lead to the resolution of formal theorem proving, without any human guidance.

The next step in the computer hardware evolution is to drastically change the way traditional computers are working: abandon the classical binary logic for a more exotic way of reasoning; this is quantum logic (we will deal with it later), but it is still far from being mastered. Maybe this could change the relationship between the computer and the mathematicians or lead to a new kind of computer which would no longer care about the mathematicians!

## Supercomputers

A supercomputer is intended to develop very-high-level computational capacity; there was from the beginning a competition between the increasing demand of performances and the also-increasing offerings of the technologies. The historic start[11] of the race was made by Seymour Cray (CDC 6600) in the 1960s, and from then on it never slowed down. The arrival of big data facilities and the Internet vigorously boosted this virtuous circle.

Now there is a fierce global competition toward the top record of the fastest machine[12]: the game unit universally accepted is no longer MIPS but FLOPS.[13]

In fact, this rough classification of supercomputers is rather esoteric, because the efficiency of these computers, even though directly dependent on the size of the machine and the number of processors, has to be strongly weighted by its structural adaptation to the issue at hand. Currently, there are three categories of "super calculations":

- High throughput computing (HTC), which implies a large amount of computing power for a rather long period;
- High performance computing (HPC), which does the same but in a much shorter time;
- Many-task Computing (MTC), which involves a multitude of separated computers related in a network for a distributed activity.

---

[11] This was quite heroic, because the transistors of the time were individual bipolar germanium devices. Today this seems rather simplistic and prehistoric!

[12] Currently the blue ribbon belongs to "Tianhe-2" (China), with 33.86 petaflops ($33.86 \cdot 10^{12}$ operations per second).

[13] MIPS: million instructions per second. FLOPS: floating point operations per second.

It is admitted that the top super machines currently have a lifetime of three years among the leaders, before being surpassed by a challenger and being in some way obsolete. Given that the corresponding budgets are especially high, this provides evidence for how much importance is placed on it by the various governments and industries around the world.

After these general purpose computers, there also appeared "special purpose" systems such as MDGRAP-3 for the calculation of protein folding[14] or the "Deep Crack" for Data Encryption Standard (DES) cipher; these specific computers use customized VLSI or FPGA chips. But, along with these frozen structures, other computers, such as the famous Watson (IBM), can possibly be largely redeployed from a task to a different one (Watson, Deep Blue).

### Supercomputers — What For?

There is a wide range of computationally intensive tasks in various fields (weather forecasting, molecular modeling, physical simulations (nuclear weapons), aerodynamics, cryptanalysis, etc.) which are relevant to super-computers. They all refer to large amounts of data coming from multiple measurements of any kind, related to industrial applications. These results are stacked in huge data banks or data centers provided with plenty of hard disks and connected to the Web.

### All-purpose supercomputers

The first job of such computers, of course, is to "crunch" the mountain of raw data, to select the relevant ones and put them in the right order before entering a program of "intelligent" interpretation. For instance, the NOAA[15] crunches billions of measurements (temperature, wind speed, humidity, clouds, and so on) every day from earthbound weather stations as well as satellites in order to make the most accurate forecast of the weather on a worldwide as well as a local scale.

---

[14] "Blue Gene: A vision for protein science using a petaflop supercomputer", F. Allen *et al.*, *IBM J.* 40(2), 310–327, 1999.

[15] National Oceanic and Atmospheric Administration.

- Healthcare also benefits from a similar procedure with the new prospect of "predictive medicine." The basic idea is that diseases (and death) arise from a lack of knowledge of our individual biological state. A better knowledge of our body should result in anticipating in due time the problems to come. Then, the more we accumulate biological data, the more we should expect to live to an old age! This procedure, if extended to a sizable population, also allows comparing the efficiency of drugs and bringing forth correlations between various medical cases to improve our knowledge.

  But, if one extends this individual philosophy to the full population of a country (let alone the full world!), I would ask you to imagine the volume of data to accumulate and perpetually stir. Thanks to supercomputers: in that special case we do wish they could be very intelligent!

  The same thing is true for many other common situations of our daily lives, but there are also other categories of big data processing which are more related to a particular scientific prospect.

- Cancer may be one of the most important applications. It is currently admitted that this uncontrolled scourge originates at the cellular level with the genetic control of protein fabrication. Basically, there never are two identical cancers (even though typical categories have already been selected); each cancer has its own behavior and each requires a particular cure to be adjusted, following the proteins involved and their metabolism. This approach is tremendously complex, considering the huge number of possible combinations of different protein molecules (some 360,000 different species have been recorded in the blood). Huge efforts are undertaken to master this especially "multidimensional" issue, and the supercomputer should be the only trustworthy approach.

- Genetics also is deeply concerned with proteins, with DNA and RNA being very complex mechanisms of production. This is strongly related to the genes, whose properties are still to be deciphered. The new science of epigenetics will assuredly bring some light to the future of this field.

- Also to be mentioned are the many endeavors to "digitalize the world" (as Google puts it) or to "understand," simulate, and possibly copy the human brain as IBM entices it with its Blue Gene/P, which aims to simulate 1% of a human cortex or a complete rat brain!

## Brain-devoted computers

Aside with these performing machines, a new class of supercomputers has emerged which tends to reproduce the behavior of the human brain but without directly copying the biological neuron. Called cognitive computers, they combine artificial intelligence with machine learning algorithms. A specific microchip called True North was developed by IBM, with a new kind of structural architecture different from the classical von Neumann one.

IBM continued in that direction of cognitive computers which use an adaptive input device inspired by Hebb's theory, instead of the usual deterministic machine language. This was instantiated by D. Mohda in the framework of a SyNAPSE program.[16]

These machines are especially devoted to situations where conflicts, ambiguity, or uncertainty dominates, which human brains encounter (and effectively resolve) frequently. The advantage of the computer over the brain is its ability to handle a much larger collection of diversified information. This is to such a decisive extent that unexpected, suggestive, or instructive ideas could be proposed as the best solution to the issue, if not the right one.

These "smart cognitive systems" infer and in some way think well beyond the limits of the basic preconfigured computing rules, and they also possibly build on the IT resources. They learn by cumulative experience and digging in big data sources. Blue Gene is a running project at IBM which aims at reaching operating speeds in the range of petaflops, with massively parallel architectures. It is intended to contribute to the particularly complex study of protein folding, which is a key element in the understanding of cerebral neurodegenerative diseases such as Alzheimer's or Parkinson's.

Neuromorphic computers compete with the biological neural system in order to lower the level of the required electrical power. They are considered to have significant promise, in an attempt to build a new kind of

---

[16]"A computer that thinks", Dharmendra Modha, *New Scientist*, November 8, 2014.

computers which may be comparable to a mammalian brain (matching a mouse brain, for instance, in size, function, and power consumption). This is quite ambitious but could be realistic. Some say a 100% humanlike scale simulation (estimated to require 4 petabytes of memory) should be achieved by 2018.

However, some opponents have published very strong criticisms of the project. Henry Makram, for example, bluntly called it "a mega public relations stunt — a clear case of scientific deception of the public"! Still difficult to say more but, ultimately, time will be the judge.

### *Supercomputers — What Are They Made Of?*

In the early years (1970, to be precise) a "super" computer involved only some individual processors, but that quickly rose to thousands (1990), to finally the fabulous level of tens of thousands of off-the-shelf multicore processors now incorporated in the parallel structure machines.

From the beginning these parallel structures were concerned with gaining speed, but they ushered in the age of massively parallel systems. The type of structure involved varied from the "grid computing" distributed approach to the "computer cluster" approach to the more recent "three-dimensional interconnects," also with "graphic processors."

Of course, the unavoidable counterpart of high density processing remains the energy consumption and, correlatively, the corresponding heat emission. For instance, the former champion Tianhe-1A consumes an electrical power of 4.04 megawatts, a huge part of it being turned into wasted heat in spite of the efforts to limit the emission at the processor level as well as the assembly level! This shows that the efficiency of the cooling systems (forced air or liquid cooling systems) became a highly limiting factor in the packing of multiple processors, and that heat management has become a major issue in high power machine conception.

Another difficulty for supercomputers to grow in size is the management of possible failures in the hardware (processors as well as interconnections). Historically, it is to be remembered that the first big computer (ENIAC) used for the A-bomb design in 1942 was so deeply concerned with this issue that naysayers thought it would never work due to the poor reliability of the electronic lamps of that epoch. They said the statistics forecast that it would be in a permanent breakdown state. They were obviously wrong, because of the redundancy of the circuits (and a huge stock of lamps!).

This problem remains valid today in spite of the extremely high reliability level reached by the current processors (which often are taken off the shelves).

As for the operating systems employed in these computers, Linux or derivatives are dominantly used, but there is no standard because of the diversity of the architectures, which are optimized for a better yield as a function of the hardware design. Dedicated software tools also exist to prevent the loss of any time in transferring data from one "node" to another. This means it becomes more and more difficult to test and debug the circuits.

## Supercomputers — Who Pays?

Now, let us take a look at the running programs and corresponding financing. There are a lot of projects which are more or less related and even sometimes co-funded in a sort of dedicated intricate research network at a global scale. Here we will try to clear up this mess.[17]

### Blue Brain Project (BBP)

This program was set up in 2005 in Lausanne (Switzerland) at the Ecole Polytechnique Fédérale de Lausanne (EPFL) to develop a "whole brain simulation" and "mind transfer" copying through a reverse engineering procedure.

Henry Makram is still in charge of the direction of the project, and he brought together an impressive multidisciplinary team of researchers. To date, a neocortical column equivalent of a rat has been virtually completed.

---

[17] In December 2015, a nonprofit society (OpenAI Inc.) was created with a mere US$1 billion funding from Google, Facebook, IBM, and others, with the aim of developing brain simulation intelligence in open source.

This was done under the sponsorship of IBM, with the complementary aim of seizing this opportunity to check the behavior of the Blue Gene computer and its successive updated versions. The main financing initially came from the Swiss government and then from the European Commission, for a total of 1.19 billion euros over 10 years. The only electrical consumption of the computer amounts to US$3 million a year, which makes it a non-negligible cost.

A joint effort is being made by the Technical University of Madrid (Spain) under the framework of the Cajal Blue Brain Project, which was associated with the BBP. Another co-project was also born to tackle the issue of the new cerebral therapies more precisely. It was named the Human Brain Project and was also hosted by the EPFL. This program is directed by Makram too, and is also funded at the same level of 1.19 billion euros by the European Commission, in the framework of the European flagship Future Emerging Technologies (FET). Thousands of researchers have already been hired, and are still collaborating while waiting for an adapted super supercomputer of up to 1 Exaflop,[18] which could be achieved by 2020.

IBM is also on track for a 1-exaflop computer. A dedicated project has been launched in Montpellier (France), at the local PADC[19] center.

Also to be mentioned is the BigBrain project of some Canadian and German researchers, which is now included in the BBP. A 3D model of the neural connections in a brain has been obtained from a microtome exploration of a brain. However, the coherence and completeness of the mockup is not yet satisfactory and the microscopic resolution does not reach the much lower neuron scale.

## Brain initiative

Brain Research through Advancing Innovative Neurotechnologies (BRAIN) was a collaborative "moon shot" initiative launched by President Barack Obama (2013) to support and favor collaborations for developing the understanding of brain function. This effort is at the same level of

---

[18] "ExaFlop" means $10^{18}$ operations per second!
[19] Power Acceleration Design Center.

financing as what was previously the White House Neuroscience Initiative, which started the Human Genome Project (HGP) in 2003, with support of US$3.8 billion over five years (and meanwhile generated a return of US$796 billion; this is worth mentioning).

The Blueprint for Neuroscience Research organizes a network for 15 institutes and partners: NIH[20] (which received US$34 million in 2014), NSF,[21] FDA,[22] DARPA[23] (in the framework of the SyNAPSE/ IBM project, headed by Dharmendra Modha), NITRC[24] (cloud-based environment for a large volume of data), and IARPA[25] (cognition and computation), etc.

The aim of BRAIN is to promote a dynamic understanding of the human brain in order to find solutions for neural diseases like Alzheimer's, Parkinson's, depression, or traumatic brain injuries. The US$100 million immediate commitment will be part of a US$4.5 billion engagement over 10 years.

This avalanche of billions is also complemented by a private sector contribution which is also quite awesome:

- US$30 million from US Photonics Industries
- US$5 million from Glaxosmithkline
- US$65 million from the University of California (Berkeley)
- US$40 million from Carnegie Mellon University
- US$60 million from the Simons Foundation
- US$60 million (a year) from the Allen Institute for Brain Science
- US$70 million from Howard Hughes
- US$40 million (a year, over 10 years) from the Kavli Foundation

With all of that put together, there is no doubt that things will evolve rather quickly!

---

[20] National Institutes of Health.

[21] National Science Foundation.

[22] Food and Drugs Administration.

[23] Defense Advanced Research Project Agency.

[24] Neuroimaging Informatics Tools and Resources Clearinghouse.

[25] Intelligence Advanced Research Project Activity.

## Google brain

This section dealing with supercomputers cannot be completed without saving some words for Google, which has a rather special place in the world of supercomputers. Google in 2013 already implemented more than two million servers in data centers distributed worldwide, aiming to soon reach a total of 10 million. This awesome computational potential of the "Google platform" makes it, *de facto*, the most powerful assembly of computers in the world, and in the mind of Google it is a super supercomputer.

In 2012, a cluster of 16,000 computers successfully mimicked a part of the mind's activity — enough to process an image recognition configuration able to recognize a cat's face among plenty of different images. Google is deeply involved in AI — the "deep learning" and the "self-taught learning". This made it possible to bring about the simulation of a "new-born brain," as the company put it. In 2014, it bought the company Deep Mind Technology, without revealing the amount for the acquisition.

However, it is to be kept in mind that if the human brain is evidently involved in the model of the conception of the circuits, it could not be considered, literally speaking, as a proper "copy" of the brain. As Yann Le Cun,[26] puts it: "While Deep Learning gets an inspiration from biology, it's very far from what a brain actually does."

Of course, when one is dealing with Google, it is not possible to speak of an involved budget (which can only be enormous), because the whole of its activity is devoted to the AI "artificial brain" and the details of the financing are unknown. Somebody said: "Google is not a research company; it is a machine learning company."

The whole strategy of Google is to digitalize and memorize everything in order to put this knowledge at the disposal of a future artificial (human-like?) brain. I could have said "brains," because if a computer is able to manage a brain nothing will prevent it from managing a collection of brains.

---

[26] Le Cun is the (French-born) AI director at Facebook. He remains famous for publicly warning his counterpart Ray Kurzweil at Google about excessive media hype ("too much, to be honest," he said!).

## Baidu, the Google of China

Andrew Ng is a Singapore-born Chinese who was in charge of the "cat face" experiment at Google X Lab. He just joined the Chinese web giant Baidu, which invested US$300 million in a research center in Sunnyvale, California — with the ambition to "change the world"! But the really impressive equipment of neural networks is in Baidu's Institute of Deep Learning, close to Beijing. Of course, they did not mention the budget involved, which should also be significant.

## The Brain/MINDS project[27]

This is the first initiative of brain mapping launched by Japan (2014) on the idea of exploring the brains of Marmosets (a kind of little monkeys which are known to share many behaviors with humans). The intention is clearly oriented toward diagnostics and treatments of psychiatric and neurological disorders.

The initial budget should look rather modest at the start (about US$27 million for the first year) but it is expected to rise significantly. At the center of the organization are the RIKEN Brain Science Institute and the Keio University School of Medicine. The simulation is run on RIKEN's Fujitsu "K" Supercomputer, which was ranked at the top of the world in 2011.

It goes without saying that, in spite of the naysayers who loudly proclaim their ultimate concerns; the current dreamful promise of the AI associated with the supercomputers certainly cannot be sneezed at. In any case, nothing can stop the enormous effort devoted to this worldwide initiative.

# Toward Quantum Logic

Quantum logic[28] (QL) is a direct consequence of the very theoretical quantum theory,[29] which develops in a world of uncertainty at the subatomic physical states of the measurable "observables." There is no

---

[27] Mapping by Integrated Neurotechnologies for Disease Studies.

[28] Historically introduced in "The logic of quantum mechanics", Garrett Birkhoff and John von Neumann, *Ann math* 37(4), 823–843, 1936.

[29] See for instance: *Mathematics, Matter and Method*, Hillary Putnam, Cambridge University Press, 1976.

question of entering into such an arduous debate; but, rather, giving an overview of the way this new philosophy could lead to practical results in the world of computers.

QL is quite far from the usual binary Boolean logic of the "formal systems" and its "logical consequences" which we have been dealing with so far.

Quantum computing is a very promising field because, even though it is gruesomely complex, it could make many calculations that are not possible using classical computers, feasible. Among the prominent attractions of this method could be the open possibility of cracking elaborate cryptographic systems. But the real difficulty in going further lies in the development of the physical means to implement such complex theories.

The quantum device equivalent to a binary gate is called a "qubit," but there remains a basic technological difficulty in making it switch from the nanoworld to the macroscopic one and back, due to a phenomenon called "decoherence," fundamentally linked to the quantum nature of the process. However, unfortunately, this switch is also fundamentally required to operate the device. To get rid of this difficulty is clearly not obvious.

Some said that if a "big" quantum computer (more than "only" 300 qubits) could be implemented, it "could" assuredly simulate the whole world![30] Quite a challenge!

In fact, the quantum calculation does not apply to every kind of problem, but only those whose complexity is purely combinatory (say, bioinformatics or operational research, which suits fairly well our purposes). Two main experimental solutions are presently explored, without conclusive (and public) results; the first one proceeds from "quantum wells," and the other from photonics.

## Quantum Wells

Many situations of possibly suitable quantum wells have been emphasized in the past, such as trapped individual atoms or ions in an ultrahigh vacuum, but the most advanced prospects are concerned with superconducting tunnel junctions.

---

[30] *The Fabric of Reality*, David Deutsch, Viking Press, 1997.

These quantum circuits (or Josephson diodes) are also called super-conducting tunnel junctions because they are typically made up of two superconductors (namely niobium layers) separated by an aluminum oxide insulating layer some 2–3 nanometers thick. The operating critical temperature of the system mandatorily lies in the very low temperature range[31] of 4–5 K or even much lower (0.02 K to prevent decoherence) for the MIT system; this makes it a serious practical drawback. These low temperatures are required to keep the quantic device out of any external noise influence (one then could never know if the Schrödinger's cat is dead or still alive; but that is another story for the next book!).

Anyway, a special integrated circuit (Europa) bringing together 512 of such diodes could have been elaborated by the Canada-based company, D-Wave, which also seems to have implemented a dedicated operating system algorithm.

D-Wave would have sold these machines (D-Wave One: some US$10 million each) to Google, Lockheed Martin, and NASA, and nobody knows clearly what has happened since then, except that a team from Harvard University claimed that they had solved[32] a model of lattice protein folding using the D-Wave computer (2012).

### Photonics

Another way to touch on the area of quantum computing without the requirement of low temperature is optical computing, still in the advanced research domain.

The two requirements of a quantum system are isolation from external perturbations (as seen above), and the possibility of managing non-perturbing "entanglement gates" to interact with them. Photons are very promising in this field for their ability to meet both of these criteria. In many applications that makes them the only choice.

Of course, when one is dealing with photons (a particular type of boson), it is no longer a matter of MHz or GHz but light speed and optical

---

[31] The Kelvin scale of temperature (K) begins with absolute zero: −273°C.
[32] "D-Wave quantum computer solves protein folding problem", Geoffrey Brumfiel, *Nature.com*, August 17, 2012.

frequencies, also with their peculiar and essential property of retaining very well quantum correlation (if not absorbed); so they are currently used in quantum cryptography applications, for instance.

The only present limitations, compared to the existing technologies, are to be found in the weak achievable complexity (the number of photons involved and the number of interactions between them). Then the inherent difficulties of the process are:

- To store the photons for a long-enough time;
- To manipulate single photons (random emission, difficulty of interaction, lifetime, etc.);
- To control the gates in a non-random way.

Despite these difficulties, huge efforts are being undertaken to outperform classical computers. Photons do present overriding advantages, such as the speed and the ability to follow all possible paths simultaneously (superposition).

Encouraging results have already been obtained recently at MIT on the Boson-Sampling Computer. All of that has obviously been rendered possible by the technical advances in the femtosecond laser domain. Assembling a whole bunch of optical devices on a chip of silicon has also been experimented with and using dedicated nano photonic elements including waveguides. This method looks very promising.

## The Compatible Artificial Logic

Dedicated circuits, probes, or implants have been proposed for years to stimulate, bypass, copy, or monitor local parts of the brain with the help of "artificial" logic.

### Brain/Computer Interfaces

Brain implants have been proposed for a long (?) time to cure severe diseases but access to the critical point in the skull is not so easy, provided that this point is conveniently identified and localized. These implants are, however, classically used for the compensation of Parkinson's disease by local electrical stimulation or by optical stimulation through optical fibers. The effect is immediate and stunning. However, the operation is especially delicate, and perturbing side effects occur in many cases. It is to be recalled that there are never two identical brains and each causes its own issues. However a huge funding coming from Elon Musk is now devoted to implement super thin polymer fibers, which could change the game.

To use brain implants in order to enhance brain performances would lead to the same difficulties related to the non-quiet cohabitation between the delicate biological tissues of the brain and the mineral nature of the integrated circuits. The presence of foreign objects in the brain remains a source of concern and it is better to switch (if possible) to other solutions, such as external means and remote communication. This would usually be more simple and effective.

Another point is the question of possible direct communication between two brains, which is sometime raised; some call it telepathy in its natural version, but it can hardly be considered reliable and operational! A digital version should certainly be preferred when possible. This is not so easy to do, because the two brains considered are dramatically different in their constitutions; the information is in no way standard, each one processes as it can and the stored information lies in a "diluted" form in different regions of the brains. There is no possible direct communication or transfer, and a common transcoding would have to be invented from scratch. Up to now, the only effective standard transcoding, which we know well and use, remains ... language!

Maybe an adapted solution will be found with genetic selection or "improvement" (!), but this is quite another story.

Even then, external capture and analysis of the brain waves is in constant progress to substitute an external circuit for a muscular or nervous failure. More and more advanced AI improvements should assuredly lead to direct and sophisticated communication with the computer. Google talks about "mastering the brain" with neuroprostheses, whilst avoiding the eventuality of a "neurodictature," it says. "Don't be evil," it also says to its engineers!

## Chips for Brain Simulation

True North is an IBM project to copy a given function of the brain in a dedicated chip. A cognitive architecture has been raised which mimics the brain's organization — neurons (processors), synapses (memory), and axons (links); this chip operates in an "event-driven" fashion which limits the power consumption to 70mW. This supercomputer neuromorphic system, the size of a postage stamp, assembles 5.4 billon transistors, equivalent to 256 million synapses, for 46 billion operations per second.

A scaled structure of 16 chips has already been made, corresponding to 16 million equivalent neurons. This demonstrates the possibility of achieving larger implementations. We are still far from the capacity of an actual brain (100 billion neurons), but it has already taken a large step ahead.

Such a chip could also serve as an implanted prosthesis for brain stimulation or complementation. Of course, applications other than brain imitation are already emphasized in the military domain (DARPA), where AI is of critical importance.

Computers have already challenged the human brain in some specific situations; some of them were very recent and impressive. I will just recall three of them.

- After many well-known successes of the computer in chess, the story continued in 2011 with Jeopardy,[33] a game of questioning (similar to the French "Question pour un Champion") where the certified champions faced the IBM computer Watson. The computer was easily victorious and earned the jackpot (US$1m).
- But the most famous and stunning recent challenge was in March 2016 between the computer (Alpha-Go from Google) and the Korean world champion Lee Se Dol[34] of the Chinese game Go. This game is well known to be especially complex and intuitive. Again the computer was an easy winner. In 2014 Nick Bostrom, one of the leaders in the "buzz" about AI threats, anticipated such a dominance but over a time scale of, at least, 10 years! Less than two years was enough.
- Last but not least: a "top-gun" pilot recently faced a computer (the size of a smartphone) equipped with AI piloting facilities, in a flight

---

[33] https://www.youtube.com/watch?v=WFR3lOm_xhE
[34] https://deepmind.com/research/case-studies/alphago-the-story-so-far

simulator for air-to-air combat.[35] The computer was the winner in every round. The machine is able to set up strategies 250 times faster than the man's brain.

What is really stunning is that, in the last two cases, the comment of the defeated human contributors was: "I was quite surprised, throughout the competition I got the awkward feeling that the machine was aware of what I intended to do".

### Biology Turns to Electronics

A different approach to making a compatible artificial logic is to turn to biology itself — that is to say, use biological materials (compatible by nature) to make electronic equivalent circuits. That was first initiated by Leonard Adleman at the University of Southern California (1994), who used strands of DNA as a memory to store data which he was able to read afterward. Since then, the Turing machine has been proven to be constructible.

The SCRIBE[36] of Timothy Lu (MIT) has been made possible by rewriting a living cell's DNA using a flashlight in the presence of a chemical. The modification is definitely included in the molecule and can be read by sequencing it. This could really lead to "molectronics," a potential rival of classical silicon electronics.

Obviously, the response time of such devices is far longer than the milli-, micro-, or nanosecond range of classical electronics, but this drawback is compensated for by the possible access to parallel computing and the extended memory capacity. Work is in progress at IBM, Caltech, Weizmann Institute, and many other places, but these remains in the advanced research domain.

Sorry for the deeply technical or scientific details of this chapter; this could appear somewhat tough or even arduous for some readers, but how to do it differently? In this development of new digital techniques lies the origin of the challenge with the human intelligence. Each person will make his own appraisal!

---

[35] https://deepmind.com/research/case-studies/alphago-the-story-so-far
[36] Synthetic Cellular Recorders Integrating Biological Events.

# Chapter 6

# How Does the Computer's Brain Work?

The computer's brain consists of a huge diversity of machines, more or less connected, to process information in a way that we would imitating the human brain connections, but we are still far from such an ideal model even if the recent developments are impressive.

The components of this network may involve individual home personal computers (PC) or broad Data Centers. They are connected through electronic links just as the neurons of a human brain are put in a selective relation through the axons and the dendrias. But the comparison stops there.

First of all, we will deal with this network, that is to say, what we usually call Internet. This global network provides the opportunity to acquire data, make a transfer of them to another place and memorize them in a secured facility where they could be retrieved later or processed if necessary. Of course this chapter will abundantly share much of the information with the Internet which provides a bottomless source. It is observed that the references indicated in the documents displayed on the Web, seldom relate to a book; this is likely because of the swiftness required for the communication. Accordingly we shall try here to accommodate the swiftness requirements of Internet and the necessary hindsight of the books. That means we will borrow many things directly given out of the net in order to be "in phase"[1] but also add comments to derive more long term conclusions.

---

[1] Thanks to the many contributors!

## From What Comes the Internet?

Electrical communications have been introduced in the last century in the "analogic" way, mainly for the application to voice telephony. But it soon appeared that the method could be efficiently challenged by the digital one that is to say the use of electrical coded pulses. Everybody remembers the pioneering work of Samuel Morse: Save Our Souls ... — ... which saved so many people at sea.

### The Theory and the tTechno

The theoretical baseground of the digitalization was given by Nyquist and Shannon in the fifties and led to what was called Pulse Coded Communication[2] (PCM or even coded PCM). This was extended to the digital logic and circuit technology based on Boolean logic functions. This was at the origin of the computers and Internet. The discovery of the transistor effect and then the implementation of the Integrated Circuits (MOSFETs) strongly favored the development of computers and consequently the Internet.

The Internet was born initially from an idea of the Pentagon during the Cold War in 1969 in order to implement a communication standard which could be foolproof: the TCP/IP.[3] The messages are cut up into packets sent in many different directions and afterwards reassembled at the arrival. This is how the first data network was born in 1970 comprising four "nodes"[4]: its name was Arpanet.

Its main function was as a link between universities to make possible for the researchers to get real time exchanges of documents and information. This organization quickly developed into the World Wide Web (WWW) and in 1983 it was Internet which gathered all the networks in a single set obeying the HTML language and a set of pages addressable through a URL and a HTTP protocol. Internet has been opened to the trade traffic in 1990. That was, in brief, the short story of the Internet birth.

---

[2] For more details see an excellent book: *Communication Systems*, Bruce Carlson, McGraw Hill, 1975.

[3] Transmission Control Protocol/Internet Protocol.

[4] Three were located in California and the last one in Salt Lake City.

The global success was so large that a billion of connected sites were proposed in 2014 to three billions surfers. In 2018 it was revealed that the Internet uses up some 6–10% of the global electrical consumption! Nowadays 95% of the Internet international traffic is operated through undersea cables (mostly optical fibers). Huge investments come from GAFAM to serve an explosively growing demand.

## Internet, What For?

The Internet was initially a communication network similar to the telephone but potentially more versatile and with a larger capacity of transfer. It soon appeared that it could do more and was associated with large memories and specific algorithms which allowed storing the information in order to use them later. *A priori* the access to this facility is free and open to everybody but that is the visible and emerged part of the iceberg.

That did not suit people who need confidentiality (90% of businesses) and special encrypted facilities; so particular search engines were developed in a hidden Web (Deep Web, Dark Web, Opaque Web…). This is the domain of "special businesses" and traffics besides the conventional confidential exchanges required by industry, researchers, banking transactions, and so on. This somewhat looks like the deepest (or shameful) part of the subconscious of a human brain where most of the exchanges remain unrevealed and may be unholy.

Another paradigm of Internet is named "the Cloud"; it consists in using Internet to access distant software servers in order to store or exploit

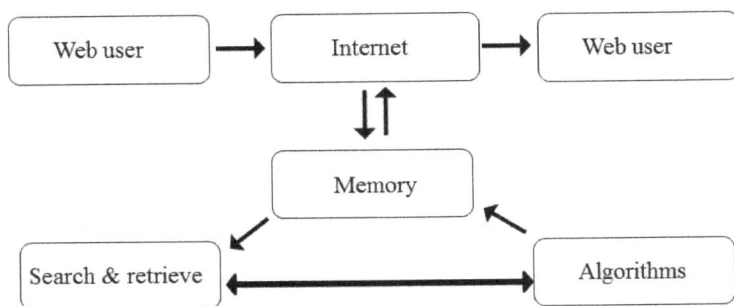

**Figure 6.1.**   The double role of Internet.

data externally. Big companies (GAFAM) have developed "cloud computing" facilities and propose to their customers, large and remote computing powers or memories (iCloud). So, step by step, the Internet is to become as multifaceted as a human brain!

The development of the Internet has brought an unprecedented disruption to our societies since the printing came along. As printing, Internet is a source of personal cultural enrichment and genuine humane development. However it also poses a direct threat to the social links if it exempts the people of shared direct relations and communication. Internet knows everything and never forgets anything.

Internet has become a partner in the current life and the introduction of the mobile phones has multiplied its impacts at an individual scale. One can say now that Internet and the machine have indisputably invaded a part of our brains with their inescapable contribution.

## Data Centers

All this activity requires huge volume of storage for this information. This is, in some way, similar to what happened with the human brains. Initially the global knowledge was shared between individual people and then had to be regularly regenerated until books and later encyclopedias were proposed to save and diffuse information. Now its volume has become so large, volatile and doomed to be shared, that the power of the computers was required to help; so, the machines were implemented that way and large centers were built to accommodate all that stuff: the Data Centers.

### Data Centers Around the World

Dedicated secured buildings were built everywhere around the world to house millions of computers[5] and keep them at work day and night in a dust free environment. The most important challenge was to reduce as much as possible the energy consumption and remove the heat produced.

---

[5]The basic units are configured as flat boxes to home the high capacity hard drives; they are jokingly called "pizza boxes".

**Figure 6.2.** Thousands of PCs!

In this view some centers were located close to power stations.[6] Microsoft even tried to immerse a center in the sea for a better cooling![7]

Data Centers consume an enormous amount of power. While some of this electrical consumption stems directly from servers' compute and storage operations, much of it also stems from Data Centers' cooling functions. It's absolutely essential for companies to keep their servers cool in order to guarantee their proper operation. This energy usage can quickly become a major financial burden. As such, any tool or techniques that can help a company improve its Data Center cooling efficiency represents immense value-add. Some also try to recover the calorific energy to use it properly or even reconvert it into electricity.

In France alone, in 2018, the required electrical power was 50% of the total power delivered by the nuclear reactors which makes a serious contribution to global warming! It is foreseen that the global electrical power required by the Data Centers might reach 8% of the global production in 2030. Iceland[8] is a very fortunate country in this respect because of its cold climate which makes the cooling of the computers units more efficient and also the geothermic sources of energy which allow an electrical production at very low cost.

---

[6] Some of them which were coal powered and have been previously decommissioned, were reactivated especially to supply a Data Center.

[7] Project Natick.

[8] Facebook already set up an important plant in Iceland.

Data Centers are interconnected through optical fibers in order to make the communication fast and "wide band". Fiber has become the de-facto transmission media across our data center infrastructures thanks to its high speed and ultra-high-density capabilities compared to its copper cousins.

They were some 4,438 centers in 2019 over the world. That means that these centers have therefore become a vital need for our 2.0 civilization and that such an artificial distributed brain is doomed to grow in the future. Of course, brain emulation is directly concerned by such a computer approach especially, for instance, in the medical field (to Alzheimer's and neural diseases, among others) based on the massive analysis of accumulated data in the Data Centers, far beyond the human brain's individual possibilities.

Computer services of the major companies (the GAFAs), as Google, are today based in dedicated or shared Data Centers. More and more companies use tools of cloud computing which stimulate the exchanges with the Data Centers and the arrival of the "Internet of Things" (IoT) will surely also contribute as well as the autonomous vehicles. "Hyperscale Data Centers" are on the road to get soon a major part of the workloads because of their augmented security, performances, flexibility, and capacity with the "cloud computing"!

### Data Centers and Artificial Intelligence

Data Centers are the global hardware for an extended AI in marketing and other fields.[9] They are the heartbeat of any IT strategy; many Data Centers were built for a simpler, less demanding age but some are silently struggling to cope with the demands of rapid artificial intelligence (AI) adoption and get their own solutions. They are increasingly harnessing AI to monitor their vast, complex network infrastructure and AI also has a crucial role to play even if Data Centers are some of the most secure places on the planet. As Gartner puts it: "over 30% of data centers that are not

---

[9] "The age of artificial intelligence arrives at data centers", Detlef Spang, August 9, 2019. Available at: https://www.datacenterdynamics.com/en/opinions/age-artificial-intelligence-arrives-data-centers/

going to implement machine learning and AI won't be economically and operationally viable by the year 2020", that's quite now! Consequently, every data-driven industry must execute AI and machine learning for their Data Centers. Moreover, AI will even aid businesses to stay ahead in this cut-throat competition. Human intelligences alone are now at the parochial level!

Global Data Centers are affected with the rise of AI in two directions: first the need for AI applications to provide the necessary computing power, second to improve the Data Centers themselves. Companies are looking into building Data Centers that cater specifically to the needs of machine learning and deep learning.

Artificial Intelligence is an invaluable weapon in the fight against cybercriminals. They can replace passive cameras with intelligent passive CCTV exploitation; also they would be able to suggest remedies to improve security. At the moment they still retain practices that seem to be from a bygone age. This required them to have engineers on constant standby that could profitably be replaced by adapted smart sensors and softwares. Running data centers and hiring staff can be immensely expensive for every organization. Furthermore, supervising and monitoring staff is an additional task which could be assisted by AI to autonomously handle various tasks like server optimization and equipment monitoring. Leveraging AI in the data center in a feasible manner is necessary for every data-driven business. Instead Data Centers that don't deploy AI and machine learning won't be operationally and economically feasible by 2020.

Data Centers are also prone to different kinds of cyber threats from cybercriminals thirsty of stealing valued information or infiltrating organizations' networks to access confidential data. Such data breaches are a common occurrence for data-driven businesses which hire cybersecurity professionals to analyze new cyber-attacks in the Data Centers and create corresponding prevention. This is extremely labor intensive and no doubt AI is of first help in this domain. Additionally, AI-based cybersecurity can screen and analyze incoming and outgoing data for security threats thoroughly thus AI systems can autonomously implement mitigation strategies to help the Data Center recover from the data outages.

AI in the Data Center can help distribute the workload across various servers with the help of predictive analytics; it can also learn from past data to distribute workload efficiently or find possible flaws in Data Centers, reduce processing times, and resolve risk factors thus maximizing the server performance and prevent equipment failures.

IT professionals need to ensure their servers are equipped to handle the increased demands generated by a wide range of emerging technologies that promise to reshape the corporate landscape in the years to come. Companies that fail to incorporate the revolutionary potential of these technologies — from cloud computing to big data to artificial intelligence (AI) — into their data center infrastructure may soon find themselves well behind the competitive curve.

AI in Data Centers will continue to consume an even bigger share of computing, network, and energy for the foreseeable future. That is the price to pay for their extended contribution. However, the advent of AI in Data Centers is showing promising potential in various industry sectors. Soon, AI will dominate the world of Data Centers and colocation service providers.

## Internet Search Engines

An Internet search engine is a software system that is designed to carry out a web search (Internet search), which means to search the World Wide Web in a systematic way for particular information specified in a textual web search query. An Internet content that is not capable of being searched by a web search engine is generally denoted as the "deep web" or "dark web".

Search engines are especially powerful in revealing hidden correlations in a mountain of noisy, low coherency, multiparametric signals which is out of the reach of a human brain. Of course this requires both a huge storage memory and a high computational power to possibly swiftly follow several tracks simultaneously.

Prior to September 1993, the World Wide Web (which was initiated in 1990) was entirely indexed by hand; webservers were hosted on the CERN webserver in Geneva. After many preliminary works (e.g. Netscape, Yahoo, Lycos ...) Google rose to prominence in 1998 and the business soon became one of the most profitable in the Internet with its algorithm called PageRank as was explained in the paper *The Anatomy of a Search Engine* written by Sergey Brin and Larry Page,[10] who later became founders of Google. In fact, Google search engine became so popular that spoof engines emerged such as Mystery Seeker. Finally Microsoft's rebranded search engine, Bing, was launched on June 1, 2009. Other algorithms have largely flourished all over the world for instance: Qwant (in French), Cliqz (in German), DuckDuckGo (in American); their results are somewhat different as those from Google because they include less advertising and invasion of privacy.

## The Search Engine Organization

In brief, when a user enters a query into a search engine it consists of a few keywords. Then the web search engines get their information by web crawling from site to site. After a certain number of pages crawled, amount of data indexed, or time spent on the website, the spider stops crawling and moves on. Some websites are crawled exhaustively, while others are crawled only partially. A query from a user can be a single word, multiple words or a sentence. Natural language queries also allow the user to type a question in the same form one would ask to a human.

The index helps find information relating to the query as quickly as possible. Some of the techniques for indexing and caching are trade secrets, whereas web crawling is a straightforward process on a systematic basis. The cached version of page stored in the search engine working memory is quickly sent to an inquirer. The usefulness of a search engine depends on the relevance of the *result set* it gives back. Most search engines employ methods to rank the results to provide the "best" results first. Most web search engines are commercial ventures supported by advertising revenue and thus some of them allow advertisers to have their listings ranked higher in search results for a fee.

As of September 2019, Google is the world's most used search engine, with the lion's market share of 92.96%, before Bing, Yahoo or Baidu. Many search engines provide customized results based on the

---

[10] http://infolab.stanford.edu/~backrub/google.html

user's activity history. This leads to an effect that has been called a "filter bubble" where algorithms strive to selectively guess what information a user would like to see, based on information about the user. That is an intelligent approach which puts the user in a state of intellectual isolation without contrary information in its own informational bubble.

## Ranking Optimization

The optimization of the ranking of a site is of prime importance as well for the user, the search engine author and the possible advertisers. Early versions of search algorithms relied on webmaster-provided information such as the keyword meta-tag which provides a guide to each page's content. Inaccurate, incomplete, and inconsistent data in meta-tags could and did cause pages to rank for irrelevant searches. By 1997, search engine designers recognized that webmasters were making efforts to rank well in their search engine, and that some webmasters were even manipulating their rankings in search results by stuffing pages with excessive or irrelevant keywords. So search engines adjusted their algorithms to prevent webmasters from manipulating rankings.

This ranking is considered as especially valuable when it ranks in the first place (of course) in the search results. These algorithms are called "search engine optimization" techniques (SEO) and they are designed to improve the visibility of a particular site. This is in contrast with the SEA (Search Engine Advertising) which aims at improving the ranking through sponsored links or advertising paid services. This competition has obviously led to hacking the search engine with an abusive "spamdexing" which deceives the referrer.

Successful search optimization for international markets may require professional translation of web pages, registration of a domain name with a top level domain in the target market, and web hosting that provides a local IP address. Otherwise, the fundamental elements of search optimization are essentially the same, regardless of language. All of that have led to crisscrossed exchanges between the machine to be optimized and the human brain to teach it how to do. No final solution currently in sight!

# Deep Learning

"Deep learning"[11] is a set of AI methods for achieving an automatic learning able to fit with the computers. They aim at achieving a high abstraction level modelling for data through articulated architectures of nonlinear transformations. Main applications were obtained in the fields of computer intelligent vision, speech recognition or natural language processing in a supervised, semi-supervised, or non-supervised way. It imitates the workings of the human brain in processing data and creating patterns for use in decision-making.

## *Machine Learning Algorithms*

Deep learning is a class of machine learning algorithms that uses multiple layers to progressively extract higher level features from the raw input.[12] For example, in image processing, lower layers may identify edges, while higher layers may identify the concepts relevant to a human such as digits or letters or faces. As early as 1995, Yann LeCun at Bell Labs tackled the problem by implementing software that roughly simulated a neuron network. This idea resulted, for a long time, in a fringe interest until recent years, when it was highly extended and achieved success in speech and face recognition. Now LeCun is head of FAIR[13] and Facebook has become deeply involved in deep learning.

Most modern deep learning models are based on artificial neural networks, specifically. They first compile the information into layers of increasing complexities, each level learns to transform its input data into a slightly more abstract and composite representation. Then the process can learn which features to optimally place in which level *on its own.*

The term "deep learning" was initially introduced to the machine learning community by Rina Dechter[14] in 1986 but other deep learning

---

[11] Also note "deep neural learning" (DNL) or "deep neural network" (DNN) or even "artificial neural network" (ANN).

[12] *Automatic Speech Recognition*, Dong Yu and Deng Li, Springer, 2015.

[13] Facebook Artificial Intelligent Research.

[14] "Learning while searching in constraint-satisfaction-problems", Rina Richter, Proceedings of the 5th National Conference on Artificial Intelligence, Philadelphia PA,

working architectures, specifically those built for computer vision, began with the Neocognitron introduced by Kunihiko Fukushima[15] in 1980. The very decisive step has been taken by Yann LeCun with his algorithm of "back propagation"[16] which had been around as the reverse mode of automatic differentiation since 1970. Typically, neurons are organized in layers. Different layers may perform different kinds of transformations on their inputs. Signals travel from the first (input), to the last (output) layer, possibly after traversing the layers multiple times. This algorithm drastically reduced the delay to process and the method has been enlarged to many situations. That shows a quality similar to that of the brain, which contrasts with the rigidity of pure logic. Also worth mentioning: the supervised learning models (support vector machines) which are used for classification and regression analysis in large datasets. Such supervised learning is the most common technique for training neural networks and decision trees. This technique is to make the machine an excellent partner of the human. As LeCun puts it: "The revolution is on the way."

Deep learning is part of state-of-the-art systems in various disciplines, particularly computer vision and automatic speech recognition (ASR). The impact of deep learning in industry began in the early 2000s, when CNN News already processed an estimated 10% to 20% of all the checks written in the U.S., according to Yann LeCun. Industrial applications of deep learning to large-scale speech recognition started around 2010. Advances in hardware have enabled renewed interest in deep learning. In 2009, Nvidia was involved in what was called the "big bang" of deep learning, as deep-learning neural networks were trained with Nvidia graphics processing units (GPUs).

While there, Andrew Ng determined that GPUs could increase the speed of deep-learning systems by about 100 times. Significant additional impacts in image or object recognition were felt from 2011 to

---

August 11–15, 1986. Available at: https://www.researchgate.net/publication/221605378_Learning_While_Searching_in_Constraint-Satisfaction-Problems

[15]"Neocognitron: A self-organizing neural network model for a mechanism of pattern recognition unaffected by shift in position", Kunihiko Fukushima, *Biological Cybernetics*, 36, 193–202, 1980.

[16]"Backpropagation Applied to Handwritten Zip Code Recognition", Y. LeCun, B. Boser, J.S. Denker, D. Henderson, R.E. Howard, W. Hubbard and L.D. Jackel, *Neural Computation*, 1, 541–551, 1989.

2012. In March 2019, Yoshua Bengio, Geoffrey Hinton, and Yann LeCun were awarded the Turing Award for conceptual and engineering breakthroughs that have made deep neural networks a critical component of computing.

## Deep Learning and Neural Networks

Artificial neural networks systems are computing systems inspired by the biological neural networks that constitute animal brains which are based on a collection of connected units called artificial neurons, (analogous to biological neurons in a biological brain). Each connection (synapse) between neurons can transmit a signal to another neuron. The receiving (postsynaptic) neuron can process the signal(s) and then signal downstream neurons connected to it. Like the neocortex, neural networks employ a hierarchy of layered filters in which each layer considers information from a prior layer (or the operating environment), and then passes its output (and possibly the original input), to other layers. Although a systematic comparison between the human brain organization and the neuronal encoding in deep networks has not yet been established, several analogies have been reported.

Such artificial systems learn (progressively improve their ability) to do tasks by considering examples, generally without task-specific programming. For example, in image recognition, they might learn to identify images that contain cats by analyzing example images that have been manually labeled as "cat" or "no cat" and using the analytic results to identify cats in other images.

The original goal of the neural network approach was to solve problems in the same way that a human brain would. As of 2017, neural networks typically have a few thousand to a few million units and millions of connections. Despite this number being several orders of magnitude less than the number of neurons on a human brain, these networks can perform many tasks at a level beyond that of humans (as playing "Go").[17]

Deep learning-based image recognition has become "superhuman", producing more accurate results than human contestants. Closely related to

---

[17] "Mastering the game of Go with deep neural networks and tree search", David Silver *et al.*, *Nature* 529, 484–489, 2016.

the progress that has been made in image recognition is the increasing application of deep learning techniques to various visual art tasks. Deep Neural Networks have proven themselves capable, for example, of a) identifying the style period of a given painting, b) Neural Style Transfer — capturing the style of a given artwork and applying it in a visually pleasing manner to an arbitrary photograph or video, and c) generating striking imagery based on random visual input fields. This intrusion of the computer in the domain of arts will be detailed in Chapter 9. This point is quite important because the arts belong to the domain of human intuition which is fundamentally not a matter of logic thus *a priori* out of sight for a computer.

As deep learning moves from the lab into the world, research and experience shows that artificial neural networks are vulnerable to hacks and deception. ANNs can however be further trained to detect attempts at deception, potentially leading attackers and defenders into an arms race similar to the kind that already defines the malware defense industry.

## Supercomputers

In the competition of the human brain with the machines a special consideration should be given to the development of super-computers which develop a high level of performance as compared to a general-purpose computer. The performance of a supercomputer is commonly measured in floating-point operations per second (FLOPS[18]) instead of million instructions per second (MIPS[19]). Since 2017, there are supercomputers which can perform over a hundred quadrillion FLOPS (petaflops). All of these world's fastest supercomputers run Linux-based operating systems and most of them display a capacity in excess of 1 exaflops. This is out of reach for human brains!

### The Competition

Today, massively parallel supercomputers with tens of thousands of off-the-shelf processors became the norm in contrast with the previous vector

---

[18] FLoating point Operations per Second.
[19] Million Instructions Per Second.

systems, which were designed to run a single stream of data as quickly as possible even if Seymour Cray argued against this, famously quipping that "If you were plowing a field, which would you rather use? Two strong oxen or 1,024 chickens?".

Over the world there are plenty of big nations which strive for having the biggest one; they are used for a wide range of computationally intensive tasks in various fields, including quantum mechanics, weather forecasting, climate research, oil and gas exploration, molecular modeling, cryptanalysis, and so on, not to mention the detonation of nuclear weapons and nuclear fusion reactors. After the U.S., China has become increasingly active in this field with Tianhe-1 using general purpose graphic processors (GPGPUs). The competition is such that high-performance computers have an expected life cycle of about three years before requiring an upgrade to keep in the run.

Software development remained a problem, but the Connection Machine series sparked off considerable research into this issue. But by the mid-1990s, general-purpose CPU performance had improved so much that a supercomputer could be built using them as individual processing units, instead of using custom chips. Systems with a massive number of processors generally take one of two paths: the grid computing approach, the processing power of many computers, organized as distributed, diverse administrative domains, is opportunistically used whereas in another approach, a large number of processors are used in proximity to each other, e.g. in a computer cluster.

More often, supercomputers are designed for a specific single problem such as astrophysics, protein structure computation, molecular dynamics or, more simply, playing chess!

Throughout the decades, the management of heat density has remained a key issue for most centralized supercomputers as it was the case for the Data Centers. The large amount of heat generated by a system may also have other effects, e.g. reducing the lifetime of other system components. A typical supercomputer consumes large amounts of electrical power, almost all of which is converted into heat, requiring cooling: e.g. Tianhe-1A consumes 4.04 megawatts (MW) of electricity. This can be a limiting factor. The cost to power and cool the system can be significant, e.g. $3.5 million per year.

## *Cloud Computing*

Aside from the traditional machines another way for heavy computing is also to be mentioned: the "networked grid computing"[20] which consists in many loosely coupled volunteer computing machines. However, basic grid and cloud computing approaches that rely on volunteer computing cannot handle traditional supercomputing tasks such as fluid dynamic simulations. As of October 2016, Great Internet Mersenne Prime Search's (GIMPS) distributed Mersenne Prime search achieved about 0.313 PFLOPS through over 1.3 million computers.

Cloud computing with its recent and rapid expansions and development has grabbed the attention of HPC users and developers in recent years. HPC users may benefit from the Cloud in different angles such as scalability, resources being on-demand, fast, and inexpensive.

Supercomputers generally aim for the maximum in *capability computing* rather than *capacity computing*. The first one is typically thought of as using the maximum computing power to solve a single large problem in the shortest amount of time. The second one in contrast, is typically thought of as using efficient cost-effective computing power to solve a few somewhat large problems or many small problems.

The run for the fastest supercomputer is open for years; after the Chinese TahihutLight in 2016 the blue ribbon now was given to The IBM "Summit" in 2018 with a record 122.3 PFLOPS. But a $10^{21}$ or one sextillion FLOPS computer is required to accomplish full weather modeling, which could cover a two-week time span accurately. Such systems might be built around 2030.

---

[20] Also denoted as "super virtual computer".

# Chapter 7

# Artificial Intelligence

Artificial intelligence was not born yesterday; it was provided as a direct consequence of the machine era. As soon as James Watt made its "regulator" the steam machine was able to take care of itself alone, and then it became in some way artificially intelligent in a specific task. Obviously, the rise of the computer era bolstered this effort in the assisted autonomy of the machines.

Before one deal with "artificial" intelligence, one should attempt to define what is to be emulated — that is to say, what "natural" intelligence could be. In no case would it be possible to copy something we do not know precisely enough, something which escapes a clear definition or observation.

## What is Intelligence?

Intelligence is deservedly a very vague notion, prominently variable, depending on the concerned individual, the moment, or the circumstances. Plato said it is "the activity which allows acquiring science"; that could seem quite restrictive, with due respect to him! He also proposed two kinds of intelligence, whereas the current cognitivists say there are nine and Wikipedia says there are 26! It could be entirely a matter of assessing the situation; let's not go into the details.

Such a definition would also be variable, depending on who defines it. Some say in short that it means to make the right decision at the right time, but this can only be known after. Then would intelligence only be intuition? Would our subconscious only be a deterministic machine with its hidden exploitation rules? A black box with only input and output? Would the choice of output follow an internal programming we call personality, character, or mood?

Geneticists believe they have demonstrated that intelligence is partly hereditary and related to specific genes (NPTN or HMGA2, for instance). Nature-and-nurture dialogue? A learned man accumulates much acquired stuff, but is that enough to really be intelligent or is it no more than an element of the intelligence? The acquired stuff includes not only the memory of the facts but also the corresponding know-how, the experience.

## What is Intelligence Made Of?

In the animal kingdom, it has often been said that man is the only species endowed with this faculty of intelligence, which puts him above other animals. Is that really true? Then my cat would be stupid? I would think not; there is in its eyes a glimmer which persuades me otherwise. If only we could teach it words! It should assuredly say many things about humans! Then, maybe, the contemptuous Anglo-Saxons would say "he" instead of "it"?

Words are at the very origin of culture,[1] which in its turn is the basis of intelligence. That makes a difference with the animal, the "language" of which is rather limited, even if it does exist: birds sing, wolves howl, the whale emits sounds which extend to incredible distances — all those signals do transport information, but is there evidence of intelligence? Even ants use chemical messages (pheromones) to carry structured information in their highly organized societies.

One may have doubts about the so-called intellectual superiority that man takes credit for, but we reassure ourselves, thinking there is everything in the human race — good as well bad (sometimes very close to animals, or even worse). However, generally, in the animal kingdom, the human brain undoubtedly represents a major achievement in the evolution

---

[1] Some say culture could be considered to be what remains after everything has been forgotten!

and biological optimization, even if this quality could sometimes be independently developed. In spite of this possibly flattering picture, the human brain and the corresponding intelligence may present weaknesses or deficiencies which we are hardly aware of.

Then what makes the main difference between humans and animals, what makes him "intelligent" is undoubtedly language. We have developed conventional identification systems (languages) which relate, in a precise and coded manner, sounds to a meaning. In addition, we actually make a visual coded correspondence with words by the means of writing, which serves as a storage read-only memory for the "live" signals of the language.

Words are similar to cerebral boxes, each related to a precise, material, descriptive, abstract, imaginative, or virtual concept. Words are the very basis of structured thinking. They are able to convey an identity, a detailed monitoring as well for things of our environment or, abstract notions. By assembling them it becomes possible to build coherent thinking. We do think with words, and a refined mastery of the language reflects a smart mind. Literary subtlety is required even for the scientific culture, because it guarantees the absolute expression of our thought, even scientific, in all its details and its necessary accuracy.

Switching from one language to another is a very difficult effort for a brain which has not been trained when young; a complete restructuring of the brain is involved, to be implemented in parallel with the original wording. Our European languages originated deeply in Greek and more generally Latin, which is at the basis of our semantics. Even a distant knowledge of these ancient languages helps considerably in recognizing the roots of the words, thus giving their exact meaning and their intelligent assembling in a sentence which our consciousness presents to us.

Unfortunately, the language we find on the Internet or social networks represents a serious degeneration of our culture. Some even think that disregarding the ancestral customs represents an intellectual breakthrough; they conclude that when they do not understand, that means the message comes from a very intelligent person whom their poor mind was not able to follow! This trend is overly emphasized when dealing with "art," a field where any rationality is vigorously disputed. Then one can raise a disturbing question: Would "art" be an expression of intelligence and to what extent? This important point will be developed in Chapter 9 dealing with AI in the field of "art".

In human collectivity, illiterates are to be considered as brain-crippled, incapable of even moderate intelligence. Some proposed the Turing test as a definition of a basic intelligence of the machine but the scope is much larger.

When speaking of man, in general, one intuitively means a kind of "averaging of the samples" which are gathered in more or less homogeneous societies. The intelligence level is rather scattered and we do not have any physical criterion that can be confidently associated with intelligence. One has to imagine cultural, vague criteria which, globally, could be a reflection, but in any case a "metrology" of the intelligence: this is the intelligence quotient, or IQ. This tentative index tries to mix and weight culture, ingenuity, and creativity with imagination, intuition, feelings, and affectivity. This is poorly quantitative and belongs to the mysterious dark part of the mind.

Another way to quantify intelligence (an old, extensively used, and well-accepted method) is the notation system, which depends as much on the candidate's abilities as on the mood of the examiner (whatever the chosen numerical or alphabetical scale). Cognitivists used to say that this method points to a "normative exploration" but in no case a "cognitive exploration." However, until now, no one has come up with anything better in spite of the significant shortcomings of the method. Consequently, one has no reliable means to measure intelligence; would it be the same with a machine which up to now is less subtle? How to forecast the adaptability, the relevance of the response of a brain to an unanticipated question?

Also, one must associate this notion of intelligence with the corresponding weaknesses, defects, deficiencies, or errors, because the brain, even if performing, is in no case perfect and its intelligence can fail and lead to blind spots or unexpected behaviors: love, hatred, emotions but at the same time courage, will or ambition, which must all be related to a precise goal. Here too, not any measurable evaluation can be imagined. How would a machine be trained that way?

Intelligence is highly diversified in its fields of application; of course, it can be purely deductive (mathematics) but also creative (inventions) or even behaviorist (say, no smoking!).

But, whatever the situation could be, intelligence is also the general way to reach the understanding, which remains the ultimate goal to regulate the demeanor. It proceeds exactly as algebra does. Three successive steps have to be followed in such an algebraic procedure:

(1) Gathering meaningful elements devoted to the problem to be solved (even if some are to be found deeply immersed in the subconscious).
(2) Formulating the relations which join together these elements into an assembly — a formal or even fuzzy equation.
(3) Solving the equation by logical means, often choosing the simplest solution from among many possible ones.

This is actually the path which intelligence follows to reach a final understanding; but to date, the computer remains unable to fulfill the two initial steps on its own, even if it brilliantly succeeds, in the last step, in solving equations numerically as well as formally. The day the machine completely "understands", man will be fully overrun. That is what is at stake today.

## Intelligence and Communication

The primary purpose of intelligence is undoubtedly communication between humans, and currently machines. That is what makes its very substance and motivation, because intelligence is cumulative with information. Words in their vocal or written form carry this information and that is the reason why words must have a precise meaning, perfectly known and identified by the correspondents in order to avoid confusion. In the Middle Ages, encyclopedias became essential for fixing the meaning of words.

Linguistics has its own logic and is an essential science for arranging and regulating the relationship between humans. Switching from one language to another became a key intellectual exercise as soon as long range contact developed between tribes, kingdoms, and then nations. Some languages, such as Greek or Latin, have succeeded to the point where they have become standards of culture, such as English, in some way today, acquiring a working language status.

Dictionaries have long been used to bridge correspondence, but changing one word for another is not enough to make a satisfying translation; the arrangement of the words in the corresponding language, i.e. the grammar, also has to be taken into account. The meaning of a word can vary greatly, depending on the way it is used (and even the way it is pronounced). All of that is rather subtle and the machine is barely capable of understanding, even if substantial progress has been made in the software.

All these concerns considered, it becomes possible to bridge the gaps between spoken and written languages. Information is then able to flow, and all the locks are removed. Today, relations are highly facilitated and the Internet is the dedicated intermediary. One can even find on the Web some dictionaries which make research easier and friendlier than the handling of a big book; but it soon appeared that going further in the assembling of words was an attainable target. For instance,[2] for a dictionary, it is very helpful to take into account how to use words by wisely choosing not the corresponding words but typical sentences related to various situations. This much richer and more effective approach has been achieved by only the active (and intelligent) participation of the computer.

The machine, in that instance, behaves as a genuine partner, paying attention to the human requests. Thus far and as for the research engines, the process uses keywords typed on a keyboard, but soon the voice will do the job more easily. It has become more and more important to automatically be able to read documents and make a vocal synthesis; this is where the computer and its artificial intelligence have to play their decisive role.

### Distributed and Cumulative Intelligence

In the manner of the three little pigs of Walt Disney, many distinguished people, from Stephen Hawking and Bill Gates to Elon Musk and Nick Bostrom, used to sing: "Who is afraid of the big bad computer!" To my knowledge, they all missed a key point: human intelligence has an important advantage over the computer — it consists in a huge network which is at the same time distributive and cumulative.

---

[2]http://www.linguee.com/english-french/search?source=auto&query=LINGUEE. I am greatly indebted to them.

The great technological achievements are the result of communities of brains (and hands) that worked in parallel in a somewhat synchronized and planned manner. This means that making an Airbus 380 or landing a robot on a comet requires thousands of differently trained brains collaborating on a common task, each one in its own field. To do the same by itself, a computer would then have to assemble plenty of dedicated processors and humanlike robots with an adapted education. To win the match and reach human equivalence, the computer would have to gather the individual intelligence of 7 billion brains, because it will be in competition with this many brains (even if not every one of them performs satisfactorily).

That's the reason why I could have better titled this book thus: "Brain(s) vs Computer(s)"! In that way of distributed collective human intelligence, Google, to be competitive with its "super computer," would have to gather not millions of computers but likely 7 billion of them! That is a lot more!

For the sake of completeness, it is worth mentioning that human intelligence, though individual as well as collective, is nevertheless cumulative. The major scientific discoveries which over the ages look individual (Archimedes, Newton, Pasteur, and even Einstein) are currently necessarily collective, through teams of involved researchers and corresponding performing instruments to be implemented.

In any case, even these individual discoverers are only the result of a culture accumulated over a very long time also. Andrew Wiles, after a long and weary tour, reached in 1994 the summit of a mathematical Everest: Fermat's last theorem. Undeniably, he was the first to place a flag on the top of this mountain of intelligence. He was alone in getting the winning idea, but lots of other bright climbers (and sherpas) had paved the way for him. Would computers have been able to perform that exploit of pure accumulated intelligence ... and intuition?

Intelligence is actually very diversely distributed among humans; some have large capacities, and many not so large. Nevertheless, intelligence may adopt very different forms. The purely intellectual domain is in no way the only one involved. Intelligence can be practical as well: I know intelligent people who are unable to correctly hit a nail on the head; others who are not as clever are capable of outstanding artistic or technical improvisations.

Intelligence is not only pure abstraction; there is not only Einstein, Pascal, Leibniz, or Newton but also Steve Jobs, William Shockley, or Linus Pauling, to name just a few famous people. In human life, this kind of intelligence does not always matter; there are other aspects of intelligence which are also valuable — judgment, will, and even emotions. No mention of wisdom, a notion which (at the moment) remains out of the reach of the computer.

Intelligence is often bound to appreciate unexpected situations in order to design an appropriate response with efficiency and advisability. The pilot who crashed his plane into the Hudson River on January 15, 2009, was not a Nobel Prize or Fields Medal winner, but he showed in the circumstances a rare intelligence and professional know-how, as every passenger can attest.

If AI is to identically copy human intelligence, it should certainly have to enter into all the details of the many compartments of man's intelligence. Cleverness also means intelligence, and demeanor belongs to intelligence (or would). It would also be required to choose the person to be copied from the vast diversity of humans — that should be a bit tricky! Maybe it would be better not to copy a single brain but a lot of brains in order that a convenient choice could be made in each case. Or they could be mixed to keep the better of each![3]

Then, before dealing with AI, let us start attempting to get closer to the simple intelligence. But this notion escapes our intelligence — this is a paradoxical situation! It nevertheless remains that, in well-restricted domains (which are, maybe, rather marginal) of human intelligence, the computer is already far ahead of the human brain.

Even if AI is not exactly concerned, there is a clear limit to how the computer behaves as an unrivaled champion; that is when deep calculations or a large amount of data is involved — the prediction of weather, the gathering of information and corresponding research (Internet), the regulation of air transportation, banking exchanges, part of healthcare, genetic analysis of the ADN molecule, and so on. These domains are kept well beyond the control of the human brain, and this superiority was highly recognized from the early times of computers — that is to say, the

---

[3] Provided that we previously got a clear definition of what is better and what is not.

ENIAC (Electronic Numerical Integrator and Computer) in 1944. It is worth mentioning that no one has ever attempted to check the results by hand! We willingly trust in the machine!

# What is AI?

Since the initial tentative research into AI around the year 1956, the evolution of the field has been chaotic; enthusiasm and pessimism alternated as computer technology emerged and improved, following Moore's law. By now many believe in rapid and impressive success; others are more skeptical. Nevertheless, computers are actually able to perform jobs which humans are unable to do, and conversely humans can easily perform tasks which a computer is (yet?) unable to do. This is the present situation, but changes could come very soon.

## *Compartments of AI*

Intelligence is an entire thing, but people like to have separate compartments when dealing with artificial intelligence.

- *General intelligence* or *strong AI*
  Also known as *artificial general intelligence*, AGI (matching human intelligence), is the simplest category of AI, emphasized by Nick Bostrom. This consists of teaching the machine how to learn, to reason (to some extent), and to plan out. This obviously does not cover the whole field of intelligence; there must be added a mind and a consciousness, which also implies self-awareness and own character (not so obvious to simulate realistically).[4] Reverse engineering of the brain[5] may be a way to move forward.

  Successes and setbacks followed, and it is recognized that the difficulties were more stringent than anticipated. Nils Nilsson[6] foresees that 20 years is required to accumulate a variety of performances which

---

[4]"The singularity isn't near", *MIT Tech*, Paul Allen, September 17, 2014.
[5]Hawkins, *On Intelligence*.
[6]*Artificial Intelligence: A New Synthesis*, Nils Nilsson, Morgan Kaufman, 1998.

could mechanize human level intelligence. A poll of AI experts found that 90% felt that human level machine intelligence (HLMI) will have been developed by the year 2075, in any case within this century.

Ray Kurzweil seems very optimistic about a possible fast development of such a synthetic intelligence reaching a human level or even higher, before the year 2045. Many other researchers in the field of AGI are more skeptical.

- *Super artificial intelligence*

A step further is super artificial intelligence (SAI), which could be reached pretty rapidly after HMLI is achieved, because the main obstacles have supposedly been overcome.

This will lead either to unexpected, very good things or to disasters (or both); there will not be an in-between. The underlying potential of the machines is expected to be much higher than the human limits in speed and transfer ability. Many tests have been proposed for qualifying SAI, which is intended to surpass human limits in any domain, including the ability to learn, "think," and reason.

Would it be required to simply set up a single super intelligence, or would it be better to organize several of them which individually would be qualified in separated and diversified domains and connect them in a dedicated network? The last point is quite evident in the technical domain. However, the acquired human intelligence often proceeds from experiences completed and analyzed following subjective criteria. Would the machine be capable of such a performance?

- *Narrow (or weak, or applied) intelligence*

This is limited to well-defined tasks even if a variety of them could be devoted to providing aid to humans, though it does not attempt to simulate the whole range of human cognitive abilities.

Computers work in an exhaustive procedure; nothing is left aside, nothing forgotten. This, sometimes, is a limiting and time-consuming factor, but it also is a guarantee not to leave aside any element. The structure can be parallel or imply a set of complementary associated systems.

The main obstacle to be overcome could arise from uncertainty, weak reliance, data scarcity, and hardware limitation of capacity and processor speed. Japan launched in 1980 the "fifth generation computer system project" with a massively parallel architecture. This led to expert systems and neural networks.

John Haugeland[7] gave the name "Good Old-Fashioned AI" (GOFAI or GOFAIR when applied to robotics) to symbolic AI, which corresponds to a high level formal representation of problems. This philosophy starts from the premise that many aspects of intelligence can be achieved by the manipulation of symbols. Expert systems, for instance, so use a network of production rules, with human-readable symbols. It is now, however, admitted that this is a rather fragile process, often leading to nonsense in the end.

- *Friendly AI*
Eliezer Yudkowsky preaches that the essential point remains to protect humanity from any danger related to an overly effective AI, such as that proposed by the Machine Intelligence Research Institute, or MIRI (Ray Kurzweil, Nick Bostrom, Aubrey de Grey, Peter Thiel). In that way a "friendly AI" would be preferred whose primary concern would be not to harm or even be unpleasant to humans. Then could the computer actually become a friend? Just beware of false friends; could such a humanlike intelligence also be able to be shrewd or even cunning?

- *Seed AI*
A "child machine" was initially imagined by Alan Turing in the 1950s, with the aim of making the machine able to self- educate and obtain its own optimized architecture through a deep learning process. This could be obtained in several steps: trial and error, information, external assistance. From minimum software at the beginning the machine would be able to build by itself a personalized intelligence following the constraints and the stimulations provided by the external world. This is exactly how a baby grows, but much faster in this case. Obviously, this supposes that the machine would be in a permanent relation with the external world through a dedicated set of sensors, as the analogy with the baby necessitates. It is expected that such a machine, in the end, would be able to improve itself through recursive "self-improvement" toward a perfection which is not reachable by human intelligence.

---

[7] *Artificial Intelligence: The Very Idea*, John Haugeland, MIT Press, 1985.

Such a machine would result from rational prerequisite procedures of blind evolution, which (as is the case for a human baby) do not guarantee resulting in a "good mind" or something quite deviant. This could give rise to an "alien" which may develop goals of its own, different to those of a "good" human. The divergences of such a feedback system would have to be carefully monitored.

Obviously, the better we understand the implied learning rules for a baby, the faster we get to an operational solution. All of that being very mindful of the possible collateral damage; this is quite a pure computer game! A similar procedure has already been followed with the neural networks using psychological theories and our experience with "reinforcement learning" on animals. So, an authentic alien would be obtained which (who?) could have goals, interests, behaviors, or competences initially unintended.

However, even when we are bringing the most sophisticated electronics into the sensors, there will still be something a machine will never be able to simulate; I mean the kindness, not to say love, a mother is able to give to her baby when singing any lullaby!

To conclude about these many confusing compartments of intelligence tentatively separated, it must be ascertained that intelligence is tremendously complex, and getting an artificial one which could conform to the model would certainly not be an easy task, even if we cut the model into slices!

Some are trying to reconstruct an artificial brain from scratch — that is to say, fabricate neuronlike circuits (or software functions) which could be able to exactly simulate the behaviors of a biological neuron. This is not so easy, because we still don't know exactly how this works in detail and, on top of that, neurons are not identical and their constitution and behavior change, depending on the function they are devoted to.

However software imitations as close as possible to neurons were achieved (they are tentatively called "formal neurons") and assembled into networks (tentatively called "neuronal networks"). To tell the truth, this is not as foolish as it might appear at first glance, because, most of the time, they are not intended to exactly replace a biological assembly but only to manage a well-defined "brain-like" function in the machine "in the manner of a theoretical neuron." This is assuredly perfectly manageable even if the simulation is not exact. These circuits are configured with learning algorithms which more or less give a satisfying diversity compatible with AI operations.

Alas, the computer has a fantastic advantage which humans can never enjoy. It is able to reverse the course of its internal clock and go back to the previous state. Yes! Just press the key "Undo"! Where is the key "Undo" in a brain? Maybe the reason is that the computer is a dumb machine which does not contribute *per se* to the external life; it just depends on a keyboard. Could the computer be deprived of the key "Undo" if it becomes a "man equivalent," born to life?

## Robotics

A special place is to be given to robotics, because it was at the origin of the very idea of implementing an artificial intelligence. It was a direct consequence of James Watt's regulator and the family of machine automatisms. Here too, the field of functions and applications is immense, but the intent is the same: giving an (limited or dedicated) intelligence to a stupid machine. The introduction of the computer changed the rules; it was no longer a matter of limited, bulky hardware but flexible software dependent only on man's creativity.

As of now, however, these robots are all supervised by a human brain; even those which are said to be "at liberty" are in fact more or less tethered to a control.

Extremely diversified fields of applications are involved. Let us suggest some of them ... without claiming to be exhaustive but only illustrative:

- Industry and workshops were the first applications of AI. Countless robots assemble or paint carefully the cars or any other objects on production lines. They have been followed by even more efficient and adaptive systems called "expert systems." Self-driving cars have now become ordinary but have yet to go further; self-driving semi-trucks have begun testing on the roads of Nevada. This represents a serious threat to the jobs of the 2.9 million American truck drivers!

- Medicine: Robots have entered operating rooms and are performing delicate interventions with greater precision than a human hand. Advanced computers such as the famous IBM Watson even participate now in giving an "intelligent" diagnostic, while in Berkeley[8] an autonomous robot has been trained to identify and cut away cancerous tissues, entirely on its own.

- Military: Of course, in this field, the human imagination is boundless and improves constantly. Automatic guiding of weapons, remote piloting of planes like autonomous drones, strange "creatures" able to carry heavy loads over rough ground or detect danger (mines), and so on.

- Heavy or dangerous duties: Exploration of sites inaccessible to or dangerous for human, such as the deep sea or radioactive areas. In such situations robots do have to develop agility, strength, and flexibility equivalent to those of humans, and this is quite challenging in many circumstances.

- Androids are the ultimate goal in identifying robots with man. The Japanese are the champions in this domain, with humanlike machines able to talk, to smile, to observe as a man (or, rather, a charming woman) would. Aiko Chihira warmly welcomes visitors at the Mitsukoshi store in Tokyo; you can speak freely with her (it?) on diverse subjects and she assuredly will answer correctly ... until you are disappointingly told |she is an android. Currently, a large industrial target for the Japanese industry is to create such autonomous droids to take care of the elderly at home.

- Space also is largely open to intelligent systems for exploring and transmitting information on planets we cannot go to. Space robots have to be (at least) as intelligent as humans in order to face, by definition, unforeseeable situations.

AI-based preprogrammed behavior bots would be modeled after the human's systems: in case of a surprising event, the bot has to react and to remember the context for later use (vigilance). AI-driven robots currently

---

[8] "Humans are underrated", Geoff Colvin, *Fortune*, August 1, 2015.

participate in any human activity — even at home to intelligently guide the vacuum cleaner towards the corners! Knigh scope's K5, for its part, is trained to watch and analyze its environment in order to recognize persons and potentially call the police!

But the most lingering question in our minds remains this: Would a robot, one day, be capable of acting with consciousness as a human would? This worrying question could have received, this week (2015), a preliminary answer by Selmer Bringsjord at the Tokyo RO-MAN conference. He claimed that a robot had been effectively demonstrated to have fulfilled a "self-awareness test" for the first time.[9]

Three famous French "Nao" robots were identically equipped with a sophisticated AI. They are individually able to hear, speak, and "think" (?) similarly to a human. For the purpose of the experiment, two of them were "given a dumb pill" — that is to say, their "speak circuit" was turned off. Then they were all asked: "Who can speak?" Obviously, the three robots heard the question, and tried to say "I don't know!" but only one succeeded. It did recognize its own voice and shouted "Sorry, I know, now!" So, the three robots got the answer (the dumb two also understood they were dumb) and understood the situation.

This logic puzzle is considered a valuable test of self-awareness, leading now to machines that can understand their role in society.

Some are afraid that super-intelligent artificial "entities" will take humans' place, reducing them to the level of slaves because they will always need them for dirty work. However, up to now, the reverse has occurred: robots are becoming more and more the irreplaceable slaves of humans, who want them to be more and more sophisticated, and this is not going to slow down. Meanwhile Google turns out to be, willy-nilly, increasingly curious about our intimacy!

## Deep Learning

In the domain of back-and-forth communication with the machine, many efforts have been undertaken for years and in various directions.

---

[9] "A robot passes a self-awareness test for first time, Niamh Harris, *News Punch*, July 17, 2015. Available at: http://yournewswire.com/a-robot-passes-a-self-awareness-test-for-first-time

Optical character recognition (OCR) is now abundantly used on every PC to automatically turn a typed document (scanned) into a regular file in the computer. This was in some way the easiest exercise of AI, because it requires only the identification of standard characters in black and white or similar colors laid out in a catalogue. The software does not care about the meaning of the words; it directly translates the image of a letter into a convenient line of program. One can say there is a somewhat limited level of intelligence involved; rather, there is pure determinism to identify the involved characters in a scanning operation. Nevertheless, this piece of software is very useful for the simple individual, as well as postage or parcel delivery services, for instance. A simpler variant of the OCR is the barcode scanner/reader we can find now in every drugstore.

Trickier was the problem of handwriting recognition, owing to the variability of the shapes of the writing and the requirement for a probabilistic comparative approach. There, intelligence is required through an advanced image analysis and an optimized logic of decision. In the same way, word vocal recognition and understanding is also of prime importance in getting full compliance between humans and machines.

Of course, besides the ongoing academic work, several companies are in the loop with regard to this exciting prospect, such as Google (Google Brain), Microsoft (Beijing Lab), or the Oculus Group.

As early as 1995, Yann LeCun at Bell Labs tackled the problem by implementing software that roughly simulated a neuron network. This idea resulted, for a long time, in a fringe interest until recent years, when it was highly extended and achieved success in speech and face recognition. Now LeCun is head of FAIR,[10] and Facebook has become deeply involved in deep learning.

The story of the adventure of deep learning and its new prospects has been related in detail by Tom Simonite.[11] The aim is that deep learning development leads to software that not only recognizes but also understands our sentences and can give an appropriate answer like a human. This supposes that the machine will be able to make sense of the language

---

[10] Facebook Artificial Intelligence Research group.

[11] "Teaching machines to understand us", Tom Simonite, *MIT Technol. Rev.*, August 6, 2015.

and to understand what words mean. It is quite a step further than the bright demonstration of Watson in "Jeopardy" or the SIRI language, which answered questions by a "simple" optimized correspondence of words.

The methods implemented with deep learning techniques apply not only to words but also to any problem related to any kind of decision process, especially those related to "games" such as those which do not include hidden information. As stated by Jeff Bradberry: "Because everything in this type of game is fully determined, a 'tree' can, in theory, be constructed to a win or a loss for one of the players." Then, without entering into the details of such unsupervised mechanisms, an algorithm of heuristic search and choice on the tree can be elaborated at each level, with matching of the conflicting goals of the players. This technique is called Monte Carlo Tree Search (or MCTS) and uses statistics for each possible move in order to select the most promising one. Then many "playouts" can be played to the very end and the final result, and used to "weight" the nodes of the tree so that better nodes will be chosen in future playouts. Rounds are repeated until the move with the best evaluation can be selected. This way of doing could, at first sight, be considered very time-consuming, but computers are currently so fast!

Today, deep learning networks use millions of simulated neurons distributed in a hierarchy of multiple "layers," which allows the tackling of complex problems. For years, the main difficulty lay in the excessive number of reachable layers, but LeCun imagined a "backpropagation" algorithm which alleviated this impediment. The initial layers of the neuronal network are able to optimize themselves in order to find the simplest features. Then multilayered neural networks can learn a larger domain of functions. This shows a quality similar to that of the brain, which contrasts with the rigidity of pure logic.

The deep-learning-trained Google computer Alpha-Go soundly defeated (by 5–0) the European champion in the game of Go. Today (March 10, 2016), after the second round, the Korean Lee Sedol, the current world champion, has been as easily beaten after a five-hour competition.[12] The exhausted loser said as a conclusion: "He [it] made moves any

---

[12] It is to be noted that Nick Bostrom predicted in 2014 that such a victory of the machine should not be expected before 2024! See Bostrom, *Superintelligence*.

human could never do! I am stunned by the result!" This is a major step forward in the dominance of AI over human intelligence. An extension in the domain of decision-making strategies has now become a clear and convincing possibility.

A direct consequence of this stunning match against the machine is an impressive rise of interest in AI among the new generation of students (the millennials), who are rushing into such studies (especially in Asia).

Also worth mentioning: the supervised learning models (support vector machines) which are used for classification and regression analysis in large datasets. This is akin to having a supervisor define predetermined categorizations. From a training dataset of examples (an input vector and a desired output value), the supervised learning algorithm analyzes the training data and infers a function that can be used in examples of unseen instances in an appropriate way. Such supervised learning is the most common technique for training neural networks and decision trees (for instance, machines able to drive cars).

If fully successful, deep learning will largely outperform the requirements of the Turing test, making the machine an excellent partner of the human. Mike Schroepfer (from Facebook) said: "I think in the near term a version of that is very realizable." LeCun, for his own part, guesses this will be achieved in two, or, at the latest, five years. As he puts it: "The revolution is on the way."

Such a step forward would assuredly be fantastic, because it would open the door to a true, "cold" intelligence. The machine would be able to answer questions in a more "motivated" way; this does not, yet, actually, mean truly "understanding" the questions but we are getting seriously closer. The machine now gets the language skills and the common sense required for a basic conversation. It would then be able to appreciate, to some extent, the relevance of its answers and possibly generate comments of its own. This would open the way for smart virtual "artificial" aids (likely more reliable than a poorly motivated human secretary!).

The likely next stage of the process would be to give the machine the ability, following the human comments, to put forward its own proposals or, let us dare to say "ideas", making it enter into a self-creative approach.

Thought generally is deductive, which is within the reach of the computer, whereas the initiation of thought in a human mind still escapes analysis. This shows the necessity to fully understand the psychological and neurophysiological behavior of the human brain before trying to implement an algorithm to duplicate the process of generating new ideas. Besides the problem of speech and handwriting recognition, a parallel and similarly ambitious project is face recognition, which brings about a visual relationship with humans. This is currently under incredible development. Google, Microsoft, and many others are actively on track.

Facebook (DeepFace) has already finalized a special technique to automatically identify a person in a photo (even when seen from the back). The success rate lies in the range of 97–98%, which is equivalent to human performance. The software takes into account the hairstyle, clothes, profile, etc. Another software program, developed in Grenoble, France, by a company called Smart Me Up, allows one to determine the age, sex, and emotions (and possibly the identity, if it is related to a bank of data) in a cumulative learning process.

Still more impressive, even if anecdotal, a software program was developed using a neural network to take into account the environment (number of persons, ages, attitudes of the attendants, etc.) to generate emotional changes in the displayed portrait of Mona Lisa and in connection with its personality as it is reported in the "cloud"! Then Mona Lisa is given a sort of "life".

All of this clearly shows that AI is not only bound to solve complicated mathematical problems but also to participate in unexpected sectors which are related to intelligence in the broad sense.

### What Else?

Computer-based systems are actually able to exhibit capabilities such as computational innovations, automated reasoning, assisted decision systems, data handling, language analysis and synthesis and, of course, direct control in robotics. We are very busy teaching the machine how to understand humans, but at the same time the machine reciprocally and actively contributes to teaching humans! It is quite a feat, after all!

At the end of 2011, a computer-assisted MOOC[13] was proposed, free online, at Stanford University, headed by Sebastian Trun and Peter Norvig, and it quickly became an outstanding success. The initiative was abundantly copied all over the world; and, only one year later, we can find MOOCs on every topic and every language on the Web.[14]

With all of that stuff, how do we produce intelligence in order to grow an artificial intelligence? We have not yet gotten so far as to have an autonomous computer able to lead its own life (but we are getting gradually closer).

Neural networks can learn from experience: pattern recognition, classification of problems, and better accuracy than human experts without the need for blocking initial definitions.

But, at the same time, over the years, human intelligence itself has been regularly improving. New solutions for accelerating the process are considered (which is as frightening as the growing AI). Genetic algorithms and programming are very promising; they generate randomly assembled new candidate solutions by mutating or recombining variants, following maximum likelihood estimation. Then selection criteria make it possible, over generations, to improve the quality. The changes could then be much more rapid than the natural random selection. Careful monitoring of the evolution would nevertheless be required, to prevent deviations. This approach, however, calls for a large computing capacity and must be agreed to be a transgression of our moral convictions.

Be that as it may, the outreach opportunities of human intelligence are significant when they are collectively organized with convenience; but this depends on the problem concerned. There still are talents that can only be solitary; they can never be replaced by an addition of mediocrities. Intelligence not only consists in solving complex problems ... but,

---

[13] Massive Open Online Course.

[14] This makes me remember a funny story which is half-a-century old. I was visiting the very recently established SUNY (Albany, NY), and a U.S. colleague proudly showed me the new amphitheater equipped with every modern audio equipment and he said tongue in cheek: "I no longer need to be there physically — a tape will be enough." I answered: "That's fine! Students will not have to be there either — their mini-tape recorders will do the job!"

sometimes, inventing them. Human intelligence cannot be restricted to a unique criterion such as IQ or an exam; this is a common error.

## Dedicated Intelligent Systems

Computers already profitably replace people in some specific and advanced tasks where they were not expected to emerge so quickly. Machines are increasingly able to perform work better than we ever could, even in unexpected fields.

The prestigious news agency Associated Press released in June 2014 the news that its financial dispatches were, by now, composed by a robot, following the example of many newspapers whose sports column abstracts are already done by computers.

Even Google translates every written language for free, and Skype also the spoken language, in real time.

In 2004, it was considered that driving a car involved[15] "so complex split-second judgments that it would be extremely difficult for a computer to do the job"! Look what happened in the following ten years! Never say never! Admittedly, one can also say reciprocally that dishwashers cannot yet empty themselves — but is there a demand (and a market) for that? Maybe fast food restaurants?

Odd as it may sound, even court decisions could be associated with an intelligent and "cold-minded" system in order to prevent the irrational deviations and fluctuations of the judge's and jury's moods. But would we be ready to accept such a dehumanized procedure? Would we be ready to accept rationality and renounce bargaining?

Is there any skill a computer could not eventually acquire? Currently, robots are commonly used in the industry to paint cars, and they do it perfectly; they are taught to, each time, identically reproduce the gestures of a good painter. But what happens if a fly, by chance, lands on the body of the car? It will assuredly be painted too. This would not happen if a man was at the control. But is there any point to investing in a super-dedicated intelligent system to only detect possible flies in the painting

---

[15] *The New Division of Labor*, Frank Levy and Richard Murnane, Princeton University Press, 2005.

workshop? That story is to say that dedicated intelligent robots, most of the time, do not have to be excessively or universally intelligent but just enough to do the job, repeatedly. For the same price a worker might be more "universal," but his work might be less reliable.

There is always a balance to be respected between the higher specifications reachable and the basic requirements to only be fulfilled. The story goes that Henry Ford, one day, complained: "Why is it every time I ask for a pair of hands, they come with a brain attached?"

Of course, the greater anxiety of workers is still the same: How long could we, humans, be able to "add value" and preserve our jobs? For good or ill, would machines become just like people — only better?

## Perception Analysis

If, one day, a computer is to really complete with humans, the playground will have to be the same one which humans have used for millennia — that is, the surrounding world. Then, if the only aim of the AI is to drive humanlike robots, there is no doubt that these bots will have to be equipped with sensors (at least) equivalent to the five senses of humans in order to get the same information. That means altogether: speech (and corresponding synthesis), hearing (and voice analysis and speech recognition), and sight (and visual recognition of the objects or human faces). As a first step we could leave out smell, taste, and touch, even if they are not to be neglected.

### Vocal and Facial Recognition

We have already dealt with these artificial senses which relate the brain to the external world and make it possible to exchange information between humans. Speech is the first organized link. If the computer is to share its life with humans, it needs to conveniently detect and, to some extent, "understand" and reproduce the voice in a synthetic imitation. This was done for years using a reduced vocabulary of command orders (in a car, a GPS, or a PC, for instance), but from the year 2000 a wider range of software programs were developed. DARPA, Google, Microsoft, IBM, and others are on track to implement more compact, effective, and friendly dedicated AI using deep learning methods. The most famous software was the Apple assistant Siri (2005). Speech recognizers are now routinely used in fighter planes, in spite of the harsh environment.

Speech recognition already works pretty well, but that does not imply that the computer is actually able to recognize every sublet of the language and "understand" the speaker. However, the terrific power of analysis, memory, or relationship which the computer is bestowed with will certainly bring many surprises in this essential domain of the man–machine relationship, even though it cannot be imagined, for the moment, that a machine, living its own life, may spontaneously address a human (or even another machine!) to express its own feelings.

A large part of the communication between humans also relies on sight, which makes it possible to guess many things about the discussion partner: the glance, the facial features, the expressions, or the context, which all give information on the state of mind of the partner in question. This is an informal and wordless way of communication which could make the computer friendlier and contribute decisively to an adapted relationship with humans.

This leads to the idea that in order to be "at ease" with humans the computer has to "understand" what is to be deduced from the human behavior and what usually appears in the changes of his external appearance: friendliness, alertness, derision, disagreement, anger, etc. (You may have noted how many times the keyword "understand" was used.) This would guide the way the computer could correspondingly react after deciding on a strategy. Never forget that a computer, different from a human, must always be guided by a defined target to be reached. Rationality always remains behind the behavior of a computer which is supposed to never dream.

## The Virtual Reality to Come

Another way for the machine to interact with humans is to create its own universe and get them into it.

Virtual reality (VR) was born from video games and industrial simulators, which proceed from the same intent: present to the "player" a synthetic intelligent reality to play or to train, through a display on a video monitor. This made it necessary to implement an adapted and real time return of the commands to the machine. But now things have gotten worse — the aim is to truly put the player inside the virtual world by means of a special headset which cuts him off from the real world and makes him dive into an artificial one[16] (sight, sound, movement, and even speech recognition).

Now, this is the "final platform" for a fierce competition and a crazy activity: Google, Facebook (Oculus), Sony (Morpheus), Samsung, Microsoft (Kinect and HoloLens), Valve, and others are on track for this immersion. Games always progress, but many other applications arose which ranged up to psychiatric therapy! Some say VR is an "empathy machine" and will soon be everywhere and be set to become a new form of expression, away from real reality (RR). VR obviously shares plenty of cues with RR; Alex Kipman (Microsoft) puts it thus: "The amount of data and signal you get for free from the real world is massive." Then, would VR become just an adapted intelligent version of RR, following a willingness that remains to be defined?

## Big Data

Big data is a direct result of the increasingly extending memory of computers and the convenience of sharing information through worldwide communication. This accumulated data on a given problem made it possible to reveal hidden correlations a human mind would never have imagined. The complexity of a problem, whatever it is, always comes from its multiparametric nature but the "big" computers are not afraid of "crunching" mountains of data.

This accordingly led Larry Page to put forward the proposal that every problem, no matter how complex, should be solvable by a computer, provided that a large-enough amount of data has been collected and that a sufficiently large computing power is "projected" on this basis. Of course, adapted mathematics and algorithms are required which give a new development to the theoretical AI.

---

[16] Surprisingly, the virtual world, whatever the quickness of the displayed evolutions in the images, does not induce any secondary effects like nausea, headache, or discomfort. This is because the middle ear is not shaken and feels nothing.

## Big Data — Why and How?

This prospect opens the door for many investigations in a variety of unexpected domains previously inaccessible: weather forecasting, genomics, simulation, cancer research and therapy, and more unexpected fields like economics, banking, or ... crime and law enforcement! The big data phenomenon was, by far, the biggest software challenge of the decade.

Of course, brain emulation is directly concerned by this computer statistical approach. There is now a radically new medical approach (to Alzheimer's and neural diseases) based on the massive analysis of accumulated data, far beyond the human brain's possibilities. This poises the individual medical files to scale up substantially, integrating the whole medical history but also any side information about the patient which can largely be stored on the Web with the "complicity" of Google.

The main brake on the development of this filing[17] of the medical data arises from the non-standardized form of the documents and the reluctance of physicians to normalize their electronic medical records (EMR).

The two leaders of the data crunching (setting apart the Chinese Baidu, which is just entering into the competition) are obviously Google with its search engines implemented on the Net, and IBM with the supercomputers (and its unavoidable Watson[18]).

## Unexpected Results

Among the many applications involved in big data, let us zoom in on a particular topic of study which is now evolving quickly and which is directly connected with brain investigation: sleep. It has been known for a long time that sleep plays an important role in the brain's activity but, up to now, there has been very little scientific and meaningful information from the field of onirology and corresponding interpretations of dreams. EEG was able to reveal different rhythms of cerebral activity but not any available information related to health.

Now, things are changing with the coming of wearable, diversified tiny sensors tethered to a computer to monitor the activity of the brain as

---

[17] "Can Big Data cure cancer", Miguel Helft, *Fortune*, August 11, 2014.
[18] "IBM's massive bet on Watson", Jessi Hempel, *Fortune*, October 7, 2013.

well as the heart, the blood pressure, the movements of the body, the snooze, or the eye's movements which come with sleep. This type of study, called "polysomnography" (PSG), allows us to gather plenty of synchronous data[19] and to elaborate correlations between these different parameters and even genetic data (Big Pharma), which can be combined statistically over many subjects. This operating mode is linked to machine learning and AI to unlock the secrets of the sleeping brain. Fitbit, with 25 million active users, is definitely the leader, and its CEO, Philippe Kahn, has said: "We're going to help people live longer and better lives."

## The Dangers from AI?

What are the dangers AI could bring to humanity as a whole or as individuals? To answer this question requires turning our minds to the future, and then running the risk of falling into science fiction. We must be conscious of this issue before getting into a panic.

### Who Would Win?

Science obviously is proceeding fast, but at the same time the intelligence of humans is to also move forward. Even though the computer contributes to the technology, to date it is man that decides and produces all the materials which the computer needs to operate. To make a computer function, it is necessary to assemble electronics, electrical power, any kind of gadgets ... and for operators to provide maintenance. To make a computer active and to contribute to any physical task or application requires a dedicated interface and coupled machines. The computer, by itself, has no hands to work and to provide a concrete outcome. Machines are ultimately made by humans (even with the help of machines). It remains unthinkable that machines can be made by themselves *ab initio*.

For instance, we do know that glass bottle fabrication can be performed in an entirely automated workshop. But it nevertheless requires maintenance, electrical power delivery, and also trucks to bring the materials in and out. Things do not fall from the sky!

---

[19] "Cracking the sleep code", Jeffrey M. O'Brien, *Fortune*, July 1, 2015.

The range of primary needs is huge and *a priori* a computer should not be able to master, by itself, all the tasks for its survival, and each one is essential. That means the computer needs us and will need us for a while. Intelligence is necessary but just cannot do everything. Some dangers to humanity are more immediate and worrying — the uncontrolled rise of the world population, among others.

What would be the aim of such computer superintelligence? Changing the world? Changing the world in what manner and to what extent? Some believe that could happen, but it would certainly not be so easy to do; and, after all, would some changes not be desirable? Humph? Would we only be able to take into account what the machine could "say"? Would we be afraid it could be right?

Many perverse but realistic hypotheses of human conditioning can be formulated which do not require any super-AI. Some experiences in this way have already been performed or are still in service in the world. Current communication methods such as TV or smartphones are well suited for "soft" conditioning and mass spirit manipulations.

In this domain, some questioning paradigms can be invoked which converge on possible collective psychological conditioning:

- The ozone hole was a failure in concerning the public, because there was evidence that the hole filled itself! That does not matter; it has disappeared as quickly as it appeared and no one remembers the great hole.
- In Europe, the use of diesel fuel is also condemned as a possible source of cancer; the GMOs introduced into foods are regarded as a real threat to our health; all of that in a rising world population, The transition from "possible" to "likely" and then "real" is straightforward.
- Now, climate warming along with $CO_2$ saturation plays a similar role in convincing populations that they have to do penance for wasting the world's natural resources, for depriving the underdeveloped countries of their goods, and so on. The aim is to collectively focus guilt on people.
- Would the present anxiety about AI be a similar effort to induce a collective "virtual reality" designed to manipulate our minds, to stuff our heads? This cannot be definitively excluded.

We could assuredly be more afraid[20] of such "low intelligence" processes which tend to establish, at a world scale, a religion of regression, penance, and guilt, helped by the hype of the media. This is so worrisome and so difficult to check for simple people! Many individuals are fond of driving human mentalities — it's so funny!

### Some Gathered Personal Opinions

Below are some opinions formulated by some famous people:

- Elon Musk (cofounder of Paypal and Tesla):
  "Unless you have direct exposure to groups like DeepMind you have no idea how fast it is growing ... The risk of something seriously dangerous is in the five years timeframe."
  "If this is not the case of crying wolf about something I don't understand."
- Steve Wozniak (cofounder of Apple):
  "AI is, no doubt, the biggest existential threat."
- Stephen Hawking (physicist):
  "This is not purely science fiction; AI could spell the end of the human race. It would take off on its own and redesign itself at an ever increasing rate."
  "Will we be the Gods, will we be the family pets? Or will we be ants that get stepped on? I don't know about that."
- Bill Gates:
  "I agree with Elon Musk and some others on this and don't understand why some people are not concerned."

Some others say that "to be conscious of the dangers and be able to understand, we must be in contact with research teams like DeepMind. AI could cause robots to delete humans like spam!" This supposes that simple intelligent people (like us) would be smart enough to clearly understand, accept, and extrapolate what the super-intelligent specialists would say. Or does that mean simple intelligent people can be convinced to

---

[20] *Le Pape François danse avec les loups*, Gérard Thoris, *La Croix*, August 31, 2015.

swallow anything? Hopefully, some authors[21] are more cautious with these sad predictions. Some even (such as Selmer Bringsjord,[22] a professor at the Rensselaer Polytechnic Institute) dare to disagree on the possibility of an AI disaster and say: "The human mind will forever be superior to AI." This is quite comforting!

Eliezer Yudkowsky (cofounder of MIRI[23]) is more difficult to place. He does advocate a "friendly artificial intelligence," but has also published science fiction works. His contribution with Nick Bostrom indicates also a serious trend in philosophy.

Many of these pessimistic comments have issued from philosophers and theorists, far away from the material contingencies. Would they have to be regarded as sincere, serious, even simple fools, or is there some kind of hidden agenda?

The most airy-fairy but very widespread theory was put forward by Nick Bostrom:[24] the theory of the "paperclip maximizer," which imagines a computer whose crazy AI is filled with the set idea of manufacturing the maximum of an object (say, the paperclip) through every means it can find. This is provocative fiction. Then the whole world's activities (including human slaves) would be devoted to this sole aim.

## What Could We Conclude from That?

However, some preliminary comments and questions can also be formulated, in bulk, to balance the opinion:

- First of all, would the aim of (artificial) intelligence be changing the world? Maybe — but that up to the point of destroying it? Not sure it could, even if it wanted to!

---

[21] *Our Final Invention*, James Barrat, Thomas Dunne Books, 2013.

[22] *Superminds*, Selmer Bringsjord and M. Zenzen, Springer, 2003.

[23] Machine Intelligence Research Institute.

[24] "Ethical issues in advanced artificial intelligence", Nick Bostrom, in *Cognitive, Emotive and Ethical Aspects of Decision Making in Humans* and in *Artificial Intelligence*, Vol. 2, George Eric Lasker, Wendell Wallach and Iva Smit, eds., Institute of Advanced Studies in Systems Research and Cybernetics, 2003.

- There are already many intelligent dudes in the world — who cares? People who are expected to care about the interests of the countries seldom behave intelligently; they are better being smart and crafty enough to stay in place. Would machines be capable of craftiness? Would a robot be better at governing a country?

- Knowing and being able to do are drastically different things. If super-intelligence is devoted only to playing with mathematics, that will in no way directly change anything in our societies. Maybe the super-intelligent computer would be limited to a "consulting" role in the man-related issues; the only questioning in such a case would be to check the validity of its conclusions.

- Supposing, one day, every person could become super-intelligent (even with the help of a machine), what could be the changes? There would no longer be anyone to sweep the streets, to do the dirty and "disgraceful" work, and the whole society would clearly become unmanageable with so many intelligent people competing.

- Very naïvely: Would a machine necessarily be malicious if not taught that way?

- If superintelligence was put in a humanoid robot or cyborg, nobody would vote for it and let it have any power to take commands. Moreover, would intelligence be required for any government? If super-intelligence was to govern, would it be worse than the existing humans?

- If a symbiosis of humans and robots could be achieved, would that open up new dimensions for the human mind or would we still "be the family pets," as said by Musk?[25]

- Would a synthetic super-intelligence mandatorily be neuromorphic, following an anthropocentric bias?

- The safer haven of a human facing a machine ambition is not his intellectual superiority but rather his ability to manage simple but diversified and unexpected situations.

- The notion of singularity sounds somewhat sacred, like the Apocalypse or the Last Judgment. Would Singularians be influenced by a biblical fixation? What could remain of that if God was to disappear from human minds?

---

[25] The Future of Life Institute.

From all of that comes the result that AI is unquestionably progressing very quickly and will play an increasing role in contributing to human lives, but human intelligence is flexible enough to adapt to new contexts. It is not so certain that the challenge would have to be considered in terms of a direct conflict. However, that also means that drastic changes in our way of life certainly have to be expected.

# Chapter 8

# Toward An Intelligent Computer

Where is this artificial intelligence going to, following the constantly improving performance of computers? Undeniably, the technical advances bring the machine closer and closer to a true understanding of many problems hitherto reserved for human brains. However, there is still a way for it to go before really competing with the so effective and so universal human mind — but would that be the real stake?

To be intelligent, a computer must appreciate the external world as a human does, or it must create its own world and immerse the human inside, as we have seen in earlier chapters.

Philosophers are keen on giving their intimate opinion, on every subject, even if they have not any competence to make a sound judgment. They are meant to be free spirits and the essence of intelligence. Now, they strive to contribute to the technological and scientific debates which they know very little about but are so eager to gloss over. Their status as bright minds allows them all kinds of audacity.

However, some of them emerge from this situation even if their scientific competences are limited to the domain of AI. Among the most internationally famous of them is Pierre Teilhard de Chardin, a French Jesuit priest, archeologist, and theologist; he was a key contributor to the empowerment of the theory of evolution by the Church.

He was so embedded in this theory that he prophesied that a new step was coming in evolution due to the increasing role of technology in our lives — "the power of religion has been transferred to technology" — and he also said that "without deep changes religion will fade into the night."

His first essay[1] (1950) describes his belief in the probable coming of "ultra-humanity." He was fascinated by the emerging technology of computers and foresaw that they could lead us to a new stage of consciousness.

Unfortunately, de Chardin died in 1955, just before the explosion of modern science, and we should have actually appreciated his thinking in the light of the current world.

## Artificial Surroundings

### *Virtual and Augmented Reality*

Virtual reality has belonged to science fiction for a long time. We currently live in such a fast-changing media environment that our minds are often saturated and cause us to be more and more distant from reality, in the political and public spheres.[2]

"Reality" is just an electrical signal from the eyes or the ears which the brain is interpreting to create a virtuality which is called "the real world" and follows a model of thinking aggregated since our birth. The brain admits only what seems possible or up to the model. Then it happens that every reality is in some way virtual. The word "real" is not exactly opposed to "virtual," but more likely to "fiction."

The brain is also able to create its own virtuality, which looks quite real; this is the case for hallucinations, which can be induced by drugs or disturbances. A special mention is to be made of "out-of-body" experiences or visions related to awakening from a coma, which are pure illusions due to an uncontrolled biological functioning of the brain. Nothing to do with the surrounding physical world; this would more likely resemble a dream.

From that it turns out the brain can be fooled by images, sounds, etc. — which behave as illusions and could be accepted as they are — as a false representation of the real world. This artificial environment can now be fully supported by "immersive multimedia" on monitor screens or, better, stereoscopic displays, goggles, or headsets, not to mention

---

[1] *The Future of Man*, Pierre Teilhard de Chardin, Image Books/Doubleday, 1950.
[2] We could make reference to the film *The Matrix*.

holograms. The very high quality and flexibility now obtained in the design of digital images make it feasible to generate a perfect visual illusion. Sensory information can be added through sounds or even "haptic" systems transmitting a tactile sensation,[3] through gadgets such as "wired gloves."

Furthermore, nothing prevents this VR from being transmitted in a remote place, which makes possible a "telepresence" or a "virtual artefact." Google introduced in 2007 its "Street View," which now displays panoramic views from almost everywhere worldwide (even in a stereoscopic 3D mode until 2010). It was promptly followed by Facebook, Sony, Valve, and others, entering this limitless economic market of distant virtual reality.

Of course, the open fields of applications of virtual reality are boundless and still worth exploring: games, fiction and entertainment, training of pilots (military as well as civilian), chirurgical assistance, simulation of environments (architecture, urban design, or engine repair), even psychological therapy, etc.

Another way to proceed is to mix the reality with artificial constructions; this "augmented" reality aims at fooling the brain with complementary added information: a recent example of this application is given by "Google glasses," which establishes a visual link with the Web. Of course, the commands of the glasses do not require a keyboard! They are entered into the processor by the voice or a finger, but some improvements are under development to activate the glasses directly by the mind.[4]

However, the success of this innovation was not obvious, due to some unavoidable drawbacks and incompatibility in the context of daily social life, such as car-driving. This led Google to reconsider the project.

In spite of that, valuable situations were found which are very promising for the use of these glasses which allow getting real time information and communication during a surgical operation or other medical situations,[5] for instance. This is also very convenient for jet fighter pilots in preventing eye movements toward the instruments on the dashboard.

---

[3] Such a "force feedback" is especially useful for medical or military applications.
[4] https://en.wikipedia.org/wiki/MindRDR
[5] http://www.goglasses.fr

More recently, the unrelenting progress of the technology offered a new possibility of entering into an augmented virtual reality: optical lenses can be equipped with an internal wearable display[6] of digital images superposed on the visual field, thus directly competing with headsets and Google glasses. The question remains that, as things get increasingly effective, discriminating the virtual from the real may become quite subtle.

## Connected Objects

The extensions of AI toward our daily lives have become very invasive since the emergence of connected objects (also called the "Internet of Things" (IoT) or, more colloquially, the "Internet of Everything"). That means providing simple objects with embedded electronics, sensors, and connectivity to enable them to exchange data with an operator or other connected devices (smart objects).

From that, a highly parallel and independent life of the daily objects can be programmed with the contribution (or not) of the humans and the Web. Of course, the stated intention is to provide us with devoted help in any circumstance of life (even when we really don't need help!), but at the same time it is worth mentioning that Google (always Google) is on the lookout for anything interesting in our private activities.

A special mention must be given, by the way, to our life in a car. Extensive efforts are being carried out to make drivers more comfortable in their cars: windscreen wipers are set to work without the need for human intelligence; the car knows as well as you do when cleaning the screen becomes necessary! The GPS tells you the road you have to follow; even the car AI is able to park the vehicle better than you might. On top of that, Google (and now every carmaker) simply proposes a self-driving car! The computer has succeeded in substituting the human in an extremely diversified and unforeseeable task, and this has even been extended to a "self-piloting" plane — why not go further? I was told that a Japanese android has been created which is able to ride a motorbike safely!

---

[6]Innovega iOptics (WA), DARPA program.

Would there remain anything a computer would not be able to do in place of a human? What a nice life is waiting for us! Would we enter a world where our health would be automatically supervised and all of our wishes fulfilled by an intelligent and committed machine at home or outside?

The IoT comprise a huge diversity of objects, from heart-monitoring sensors or implants to automobiles, search and rescue, but also home smart thermostats, washers/dryers, cooking appliances, vacuum cleaners, locks, intelligent shopping systems, wearable systems, energy management, and so on; a plethora of applications which already exist or are to be invented for the purpose!

In France, Rafi Haladjian has proposed strange rabbits called "Nabaztags" or "Karots" which contribute through WiFi to domestic life: they can hear noises, wake you, read the mail, talk, obey orders, detect a presence, and manage other connected objects at home. This supposes that you have previously spent a lot of time explaining to the machine in detail what you want it to do! Of course I forgot to mention the inescapable smartphone that plays the role of the conductor of all these instruments.

The increasing complexity of managing these paraphernalia would be such that we would be eager to get a super AI which could do it in our place and anticipate everything we would need to do! We would then become not slaves but assisted humans unable to "live" by ourselves. This would be a marriage of the "born" and the "made" as Kevin Kelly[7] said. The myth of technology is appealing and the power of technology is seductive.[8] This crazy activity would obviously generate a mountain of data which would have to be screened, sorted, and recorded in a big data center for possible future use (Google still watches for opportunities to target the consumer!). We are now at the beginning of a new era of collective behavior where our individuality will be jeopardized; there is no need to copy our brain to make us e-slaves.[9] To be served by slaves (even e-ones) results in paying the price.

---

[7] *Out of Control: The New Biology of Machines, Social Systems and the Economy*, Kevin Kelly, Basic, 1994.

[8] *The Unbearable Wholeness of Being*, Ilia Delio, Orbis, 2013.

[9] "You aren't just going to lose your privacy, you're going to have to watch the very concept of privacy be rewritten under your nose." In "Say goodbye to privacy", Geoff Webb, *Wired*, February 15, 2015.

In the future, the IoT may be an open network in which autonomous intelligent entities and virtual objects will be interoperable and able to act independently following the circumstances in context-aware automation.[10] No need to say that hackers are waiting for new prey so offered.

The systems could benefit an "event-driven architecture" which would no longer be deterministic but instead be based on only the context itself and be adaptive to standards. It is emphasized to so encode 50–100 trillion objects.

Also, it is to be mentioned that young people are perfect targets through peer networking participation.

Some people are afraid of the danger of electromagnetic waves, especially from the relay transmitters which are so essential for their smartphones, which they cannot live without.

Would they also be afraid of the many WiFi transmissions they would be provided with at home, all day long? We have entered a "radio-electric lifestyle." This is even more worrying, because we still are ignorant of the exact possible effects (short, as well as long-term) of the EM waves on our body.

## *The Internet, the Games, and the Social Networks*

There is no doubt that young people are fond of all kinds of games generously provided by the Internet, especially violent ones which are presented in a convincing virtual reality. The impact on their young minds can turn out to be seriously damaging, and they get addicted to the screen and become couch potatoes. This psychic turbulence might really lead to a collective insanity, as stated by Noreen Herzfeld.[11]

But, more generally, everybody gets trapped in this shared activity of the social networks, to the point where some imagine that a "global consciousness" will appear. They identify this emergent evolution as a global version of the "noosphere" (from the Greek "νοῦς"), which was for a long time emphasized by philosophers like de Chardin or Vernadsky as the third phase of development of earth.

---

[10] "These devices may be spying on you (even in your own home)", Joseph Steinberg, *Forbes*, January 25, 2014.

[11] *Technology and Religion*, Noreen Herzfeld, Templeton, 2009.

The power of the computer to link humanity on a new level of consciousness suggests a forward movement of spiritual energy[12] and a complexification of the relationship. We currently are not so far from that, considering the power of the consensual influence of the media on our free thinking.

All of that was intuitively foreseen by Henri Bergson, who already proposed that man's evolution has to become "creative" and cannot be limited to Darwinian "natural selection." This "emergent evolution" would correspond to an increasing "complexity" (including the mind) and the emergence of a self-reflective universe. Would that be inescapable?

## Toward a Superior Intellect

### *Cognitive Computing*

Game-playing AI surpasses human champions (checkers, Scrabble, poker, chess, Go, or any TV game), as successfully shown by the TV game show *Jeopardy!*. A problem of narrow intelligence is that the computer remains unable to do anything else. AI is essentially a software problem. As soon as it works, no one mentions any longer it is AI.

Cognitive science often starts with thinking about thinking, but it also moves forward to teach the computer how to think as a human (or better, if possible). The idea is to combine AI with machine deep learning algorithms to attempt to reproduce the way a human brain works. The structures of the programs are built on the basis of "neural networks" which are striving to mimic the brain (in expectation of being able to later mimic the mind!).

Information is no longer attached to symbols but to "weights of connection" ... this is quite subtle for a newcomer! This procedure is intended to be more adapted to provide flexibility in pattern recognition or computer intelligent vision. It involves, in an Internet-like configuration, a network of "nodes" routing the messages in a parallel architecture which adapts by itself to domains of variable situations.

Robots are mainly concerned with such cognitive computing, as we will discuss later. They must be very close to a direct interaction with humans.

---

[12] Delio, *Unbearable Wholeness of Being*.

## Watson, the IBM's Star Story

The IBM story in AI begins in 1996, when Deep Blue,[13] a computer fully dedicated to chess, was defeated by the world champion Garry Kasparov; but the year after, the score was reversed. Then IBM never stopped improving its investments in AI. The newcomer, in 2011, was called "Watson" in order to make it feel more humanlike. This name refers to Thomas Watson, the founder of a society which would become IBM in 1924.

At the beginning, the machine was a rather big one, with a parallel architecture to fasten and enlarge the processing. The processing power was considerable: 90 servers each equipped with 32 cores clocked at 3.55 GHz!

This basic unit has since been significantly improved, and optimized to such an extent that, as IBM said, it is now limited to "the volume of four pizza boxes"!

This computer soundly beat its challengers in *Jeopardy!*; it was able to understand the questions, find the best answer, and reply with an artificial voice, in a pure Turing test operation, all of that faster than its human competitors. This victory was undoubtedly a decisive landmark in the man-versus-computer war.

A video of this game can be found on YouTube.[14] The last question was: "Who is Bram Stoker?" Would you have the answer? Yes, Watson had. In that TV show, the external aspect of the computer did not simulate a manikin but a simple standalone unit, because such a presentation was believed to be much more easily accepted by audiences in the U.S. (It would certainly not have been the same thing with a Japanese audience which is keen on humanoid creatures!). It was finally concluded that giving the computer the name of a human would show its commitment to rivalling humans clearly enough.

From that time on, Watson has been growing and undergoing steady improvements, to such an extent that it is now able to comprehend not

---

[13] A 700 kg, 2 m-tall electronic beast!

[14] "IBM Watson: Final Jeopardy! and the Future of Watson", Youtube, February 16, 2011. Available at: https://www.youtube.com/watch?v=ll-M7O_bRNg. The answer is: Bram Stoker was a British author famous for his novel *Dracula*.

only classified questions but also larger notions, and possibly generate concepts we could plausibly describe as "beliefs," as said by Ben Wallace-Well.[15] With the help of a special algorithm adapted to the old game Breakthrough (Atari), Watson in a couple of hours was able, by itself, to determine how to win every time. "Both mesmerizing and terrifying" commented Jaan Tallin (DeepMind), and Elon Musk (Tesla) added: "With AI we are summoning the demon."

In spite of these worries, Watson proceeds uninterrupted in its query of intelligence as it may now understand emotions (yours, not its) by analyzing images of faces and also generate hypotheses — that is to say, new ideas of its own; this means that machine creativity is on the verge of being achieved. Thus, the gap between machine and man has become so close that IBM CEO Ginni Rometti, dealing with Watson in an interview with Charlie Rose on CNN, repeatedly said "he" instead of "it"! This was a very clear sign, all the more as Rose did not react to this discrepancy!

To maintain the steadily increasing performance of Watson and its ability to freely interact with people, an effort has been put into the diversification of its skills by improving its vocabulary in many specialized domains, such as medicine, detective work, or engineering, in order for it to translate the knowledge, shortcuts, or intuitions of the specialists into free speech which everybody can follow. Step by step Watson is becoming more and more "intelligent" and, as said by Wallace-Well, "Watson is an expert that just begins to understand!"

Watson was recently able to interpret the words used by an elderly woman in a doctor's office to describe her pains. After growling for a short while, it gave a diagnosis and proposed a therapy! Nobody said whether Watson was able to take the pulse of the patient or hear her breathing. Maybe that could happen someday.

Other projects that Watson is currently working on is the theme of deep learning, with various approaches, such as Google Brain with Andrew Ng and the Chinese collaboration, or the AI Swiss lab IDSA, the aim always being to challenge human cognition plasticity. No need to "copy" a brain — just make a new one; this is a more promising solution.

---

[15] "Boyhood", Ben Wallace-Well, *New York Magazine*, May 18, 2015.

## *Big Data is Coming*

As stated earlier, the brain is overwhelmed as multiparametric complex problems arise. Search engines develop from deterministic processes and do not care so much about human intuition. Algorithms tirelessly and scrupulously crunch piles of data; nothing is left aside, nothing forgotten. At this level AI is not required, *a priori*, to be strictly neuromorphic.

Search engines are especially powerful in revealing hidden correlations in a mountain of noisy, low coherency, multiparametric signals which is out of the reach of a human brain. Of course, this research requires both a huge storage memory and a high computational power fast enough and configured in a parallel architecture in order to be able to follow several tracks simultaneously.

This is the case, for instance, with the research on the holy grail of eradicating cancer[16] (I would have said "cancers"); identifying the risks or, hopefully, eventually producing adapted cures. Such a project (CancerLinQ) is being developed by ASCO,[17] which collects data from every patient in the U.S., and puts them in a standard form in order to show a classification of patterns and typical, similar situations; it is currently known that each cancer is a special case to be identified. This allows making valuable comparisons of the efficiencies and the reactions to the treatments that lead to better chances of obtaining a positive outcome for the patient.

It was recently announced that 14 institutes in the U.S. and Canada use collaboratively the IBM Watson analytic engine to improve their treatments for cancers. This is a very promising way for an AI-driven diagnostic to speed up the process.

We have just been informed about the creation of a Watson health headquarters at Kendall Square (Cambridge, MA) — a platform integrated with Apple, specifically dedicated to helping biomedical companies manage the development of innovations in partnership with hospitals.

---

[16] *"La défaite du cancer — histoire de la fin d'une maladie,"* Laurent Alexandre, JC Lattès, 2011.

[17] American Society for Clinical Oncology.

# Coming Research Programs

### *Supercomputers — What for?*

In a cyborg view of future transhumans, it becomes more and more likely that implanted artificial organs (any kind) will prevail in the future. "Any kind," I said, but what about the ultimate organ, the brain?

Many kinds of implants are already used, bringing more and more sophisticated solutions to the stimulation or de-excitation of neurons; they do not require the use of a supercomputer. Such powerful machines are more dedicated to the exploration of the brain or even tentatively reproduce a humanlike AI, not to mention the efforts to bluntly "copy" a brain, which obviously will need a very powerful computer and a huge associated memory.

We have already mentioned the dedicated programs directed by Henry Makram and others.

A similar project is being carried out by Andrew Ng, a Chinese researcher; he was born in Singapore but got his degrees in the U.S. and became a Stanford University professor.[18] He is working on one of the more promising and contested frontiers: machine learning, which means teaching computers how to teach themselves.

To start, he benefited from 16,000 computers loaned by Google to test a neural network that could watch YouTube films in order to perform human face recognition and get knowledge of the situation. The process accumulates layers of concepts from abstract (colors, shapes, etc.) to more complex. This is a major breakthrough in AI, due to the great processing power involved.

Ng is currently in touch with Baidu, the Chinese equivalent of Google which is on the way to building a research center near Beijing (Baidu's Institute of Deep Learning). The aim is to use this AI approach to rummage through the ocean of information coming from the Chinese Web with this "world's most sophisticated neural network".[19]

---

[18] "Brain builder: creating the world's smartest artificial mind", Dan Kedmey, *Time*, October 27, 2014.
[19] 100 billion connections.

### Neuromorphism

This is mainly implemented through several heavily funded projects from DARPA (SyNAPSE[20] program) and IBM (Blue Gene).

These efforts proceed at the level of integrated circuits (HRL Laboratories' memristor technology or IBM's TrueNorth microchip), as well as that of the petaflop supercomputer. The leading idea remains to arrange a scalable structure with low power consumption. Here too, the architecture is massively parallel in order to adapt to various problems the same way a human brain would.

The IBM computer Watson was a famous winner of the game *Jeopardy!* but it still has to tackle very different complex situations such as marketing, finance, or medical diagnostics. The ability of the machine to handle massive collections of information often provides serendipity which a human brain would not have experienced. Cognitive systems move beyond the usual calculation rules and programs; they learn, remember, and adapt intentions.

This has extended to a very large scale with the project Blue Gene, which has evolved for years over several generations of machines to the petaflop range. It has been successfully applied to the intractable problem of protein folding via large scale simulation, for instance; but it is also intended to mimic parts of mammalian brains, such as the olfactory bulb. IBM for its part developed a cortical simulator which matches the scale of a cat cortex (Cat Brain project) or some 4.5% of a human cortex. The simulation, as of now, runs much slower than real time but unfailingly.

If the computer is to become intelligent and conscious as a human could be, would it have to be taught identically to a human? It would thus have to become familiar with all the weaknesses or unaccountable feelings of humans.

That is to say, would it be taught, for instance, in sex, which often motivates so many human behaviors (directly or otherwise)? The material configuration of a robot does not predispose it to such a "concept" which is so natural for all living things. How to explain and to convince a rational machine of the importance and implications of such an intangible issue? Maybe feelings or even love could be explained, but sex as a physical

---

[20] System of Neuromorphic Adaptative Plastic Scalable Electronics.

behavior? I wonder. It would certainly have to be previously taught what a man is and what a woman is, and why they are so different. Not so easy to explain to a machine!

The computer, up to now, is a fully insensible machine. Not any notion of pain, joy, fun, shame, love, mercy; or mood; not any such returns from experiments — the corresponding sensors are still to be invented! Of course, it can be said that there are nevertheless humans who are also deprived of such sensitivity — but are their brains worth being copied?

I agree that, effectively, if computers will, in a remote future, be able to reproduce by a sexual process (or an equivalent), there will be a real threat to human survival! I guess it will not happen, hopefully.

Another point, which is essential, is religion; in the human mind, religion is of prime importance for all eternity, even though it is sometimes rejected in favor of a proclaimed atheism. The necessity of religion arises mainly from the dangers of life and the fear of death, which requires some healing and hope in an eschatological explanation. Again, how to explain and teach such a visceral concern to a machine? Meanwhile, the permanent advances of science, which constantly reveals hidden secrets of nature, take us further away from faith.[21]

Also, what convenient representation of God (or gods) would be chosen for a machine: Jupiter and his mythology, Yahweh, Amon, Buddha, Allah or, more simply, God as a standard? A solution could be to teach the computer about every religion in order to make it able to understand every human and consequently adapt to any circumstance. In any case, the computer could not be taught to "trust" in any virtuality.

But the question remains: Would there be any requirement of the notion of God for a machine? Sure, but not yet!

## Brain Transfer?

Why imagine a possibility of transferring a "brain" into a machine? A possible and provisional answer could be to obtain a "backup" in case anything wrong happens. Also, an ulterior motive could be to provide a means of extending life artificially.

---

[21] *Is Man to Survive Science?*, Jean-Pierre Fillard, World Scientific, 2015.

## A Copycat of the Brain?

In such an opportunity, this would inevitably imply also setting up a way back — that is to say, being able to "reinitiate" the biological brain after any malfunction, as we usually do with our PC. However, this should be much trickier with a brain! If not, the backup would strictly be of no use at all. Anyway, here, we are in the realm of pure science fiction!

Instead of repairing the failing brain, would it not be better to bluntly substitute the computer for the brain, by connecting it directly to the nerves and letting it take command of the body? In this case, the computer would have to be taught all of the functional requirements of the body (food, sleep, etc.). That is quite crazy.

Another scenario is to get a Humanlike computer brain (HLCB) as similar as possible to the biological one and let it lead its own life, which necessarily will diverge from that of the biological one except if permanent updating is managed; but what could be the exact aim of the operation? Would this HLCB be intended to survive the death of the natural brain? What for? Maybe to let the new "being" be able to transfer the human life to a virtual one? Perhaps this "intelligence" could be installed in a manikin representing the deceased, to be more credible. The Japanese would adore this!

If the aim of that is to preserve a particularly intelligent brain and let it be able to continue producing its talents in a virtual life, why not establish such a super-intelligent mind directly from scratch and imitate only the useful functions? The question is whether or not we are here in a cartoon!

Then it can be concluded that exactly copying a brain poses many problems which are not easy to solve. Maybe a better solution could be to independently build a software program imitating some aspects of the brain's functioning. Many projects tackle the subject; among the most prominent is Google Brain.

## Google Brain?

Google's AI story starts with the need for translating websites between different languages. The company turned to deep learning to improve its search engine and make it more efficient.[22] "Deep learning" was a

---

[22] "Google search will be your next brain", Steven Levy, *Wired*, January 16, 2015.

nickname for the neural network organization modeled on the human brain. To do that, the code is organized in "layers," the first one being devoted to determining a key feature of the input by intuiting complex features of the words and identifying them. At the beginning this was introduced by hand, but with the help of the technique called "backpropagation" the system learned from its mistakes and progressively became able to assign its own features in order to reach an "unsupervised" state of knowledge.

This procedure does not require one to previously get a three-dimensional model of the connections of the neurons in a human brain, or a refined knowledge of the biological laws of the functioning of individual neurons. Only knowledge of the way the functions are fulfilled is enough; the remainder is a matter of AI and algorithms.

This was done by Geoffrey Hinton, and his work was kept free to use. Microsoft Research and IBM obtained it quickly. But Google was seriously convinced that neural networks would make big breakthroughs soon. Jeff Dean and Andrew Ng followed, at the moment when the power of the computer exploded. With these new opportunities they figured out how to train really big nets, and they called the project the "Google Brain" even though the official name remained "Google's Deep Learning Project."

However, this shows that, progressively, the initial target of a search engine evolved in a larger and much more intelligent machine to copy the way a human would search.

The first success was rather strange! The machine (16,000 microprocessors) was shown some YouTube videos and asked to say what interesting and repetitive things were seen in the videos, without any more requirements. The proposed video files were all displaying various cats in different situations and the machine, by itself, recognized that the interest should be in a particular "object" which could be identified in all of the videos despite its changing aspect and which corresponded to what we know as cats. Dean said: "It basically invented the concept of cats!" It just remained for one to tell the machine that the name of the "object" was "cat," and it would never forget what a cat looks like! This was a great breakthrough, a first step in the knowledge; it however remains that the machine ignores how a cat behaves and what a cat really is. There comes the issue of consciousness.

This nevertheless opened the door to a much more ambitious use of AI through neural networks, such as a neural image caption (NIC) generator, which might be extended to a system able to generate sentences in a natural language. The system now learns concepts on its own and would be able to "know" the world just like a human.

From this point on, concern was aroused about an AI which could get out of human control to the point where it had to be constrained. So Google set up an outside board of advisors to monitor the evolution of the projects and propose warnings and solutions in due time.

So now Google, as a company, builds a brain of its own instead of hiring human brains! All of this demonstrates that brain-like machines can be emulated without needing to copy a brain, strictly speaking.

### How to Train the Computer?

Computational analysis of machine learning algorithms is accurately performed in new tasks after having experienced a learning dataset. But because training sets are finite, the future, uncertain learning models will never guarantee the performances and the bounds of the algorithms.

Different approaches, supervised or not, can usually be chosen, each devoted to solving a particular problem: decision tree learning, association rule learning, artificial neural networks, inductive logic programming, clustering, reinforcement learning, and so on.

An important field of introducing intelligence into a computer relies on image analysis and, more particularly, face and correlated pattern analysis. Some very nice and surprising results have been obtained in "image sentiment" classification.

# Would the HLCB have to be Taught

## About Every Human Behavior?

First question: Would the HLCB be compelled to be allocated a sex? It has been well established by biologists (and by our self-experience) that the biological brains of men and women typically differ in size and internal configuration[23] (even though common features can also be found!). The wiring is not exactly the same ... sorry, but nature is like that. This reality is not of prime importance in our daily lives (even though such a difference could sometimes be charming), but as long as we are about to copy a brain, it becomes essential to decide which one to emulate. I wonder why this basic and obvious issue has never been invoked (to my knowledge), even by philosophers ... or by researchers.

The way a man and a woman approach a problem often takes different paths, even if the final conclusion could be the same. Same thing if the problem is presented to an American brain or a Japanese one. Education also plays a fundamental role in the connections of the attained brain.

A possible answer would be: If, one day, we might be able to copy a brain (no matter which one), why not copy both types? Then we would be led to "sexualizing" the computer! That would become quite tricky!

More generally, human behavior is often guided by sentiments or feelings that are rather difficult to explain rationally. Given the same input data, the output obtained from a computer is usually predictable; from a brain, not necessarily.[24] Of course, affection, hatred, and even love belong to this kind inexplicable feelings. HLCBs will need to be trained to identify, understand, and appreciate these subtleties and take them into account in its own behavior if it is to participate in life as a human, and especially if they are intended to think for us as well.

Emotions, too, are well beyond the control of rationality, nevertheless, they still are part of the brain; they manage time, prioritize tasks, and recognize the tragic or affective events of life. Damasio[25] found that patients, with brain damage in the frontal lobes performed normally in intelligence tests but could no longer show emotions. Emotions are essential to cognition and thus to the survival of the "thinking body." In this domain of emotions, sexual contribution is quite essential in determining what is what. How do we keep an HLCB sensitive to such intangible, elusive, and irrational feelings?

---

[23] No interest in entering into an endless discussion on the "gender"!

[24] *The Tides of Mind: Uncovering the Spectrum of Consciousness*, David Gelernter, Liveright, 2015.

[25] *L'erreur de Descartes: la raison des émotions*, Antonio R. Damasio, JACOB, 1995.

# Part 3
# How to Make All That Stuff Hold Together

# Chapter 9

# The Challenge is Open

When dealing with Brain and Computer, there are two different fields to consider, as has been emphasized.

## What About the Arts?

The first field implements logic, rationality, computation, and proceeds from known established mechanisms. There, computer feels comfortable and often largely outperforms the best brains. Memory is the key factor to succeed and adjust data and rules. We do know that the computer memory is currently immensely large and easily accessible (Internet for instance); this constitutes its key advantage. Mathematics plays an ambiguous role, as it provides the rules to be applied but at the same time requires the invention of these rules before their implementation, starting from nothing. This intellectual effort belongs to the second field.

This second domain, then, results from instinct, intuition, spontaneity, emotion, sentiments, personality ... soul, perhaps! It is pure uncontrolled creativeness without any clear origin except that it comes from the deepest part of the brain still unexplored and unjustified: the subconscious. This field has given rise to arts, philosophy, literature, (even science!) etc...

Nobody knows exactly how new ideas can emerge from nothingness, or how to stimulate them. Nobody knows how the brain makes a judgment between a good idea and a bad one. This creativity largely varies from one brain to another and those who are the best achievers are qualified of being "gifted" as if God kept an eye on them! Of course memory plays a hidden role in this process which results from the imbedded culture, souvenirs, past experiences, and so on.

The emerging idea has followed an indecipherable trial and error process in the deep brain, to reach the estimated best solution. So, with such a fuzzy and uncertain track, the computer gets lost (at least for the time being!). This will certainly be the last refuge for the brain against the invading computer.

In order to give a qualitative appraisal of the battlefield, I have focused on a domain where the computer seems to be assuredly the most uncomfortable: the domain of the Arts which is, possibly, the most intuitive and irrational one. This stems from a notion that we can hardly get a definition as long as it is so elusive, changing with the person concerned or with the time; a notion which extends all over various domains such as painting, music, dressing, even cooking or architecture; the notion of "beauty" which could even be evoked! This is arguably the more difficult challenge for the rational computer, whereas the brain is so comfortable with it.

## Why is Art the Preferred Domain of the Brain?

The word "art" comes from the Latin root "*ars, artis*" which basically means "know-how", strictly speaking. This word gave birth to a flurry of other words such as artisan, artefact, artifice, artificial, and, at the end, artist. Fundamentally, art constitutes craftsmanship, technical learned skills, science, pragmatism, logic, but that is assuredly not enough, something else more subtle is required to close the loop.

### What about the Arts?

Then, strictly speaking, art consists for instance knowing how to make a chair, let's say a stool to sit on; this is the proper domain of the artisan and his know-how; but another notion soon comes which is typically proper to the human brain: the notion of beauty. The same artisan can be sensitive to the beauty of his work and could prefer to make a "Louis XV armchair" instead of a simple stool. This artisan has become an artist as long as he is able to develop a research to improve the aspect of his object and make it more attractive for other people.

The research for introducing beauty in the work is typically a human tendency which is, in no way, induced by a particular concern of efficiency; it only comes from reaching a pleasant appearance of the achieved work. There is no definitive rule for that, whatever the domain, only an irrational feeling. This same concern can be found in many various applications:[1] music, painting, architecture, even cooking or hairdressing as well, and this is a pure and natural concern of the human brain.

Beauty obeys variable criterions depending on the artist, on the local society and on the epoch concerned. It is not enough to create something the artist appreciates; the work has also to be appreciated by somebody else who will patronize it. Even Picasso, who was a revered painter in his time, would certainly have had not any success if born in the 15th century![2] There is no definitive rule in the arts, only preferences which can be changed freely. At its limit, if well achieved, the beautiful can even reach the top level of what we say: the sublime.

All of that is fundamentally out of a rational intelligence and only belongs on a deep instinct to express a harmony that could be shared with the people around. That keeps the game very difficult to master for a computer! Even for humans, art remains rather hard to be shared or taught and putting it in an algorithm still remains a pipe dream.

In the following we will be more especially concerned with painting which is an old and very significant art worth to be analyzed. So you will not be surprised that this chapter of the book will be illustrated by lots of images to make things clear (and pleasant)!

At all times in history, images involved quite a magic evocation, whether sculpted in stone, for a statue, carved on a metal sheet or smeared on any support with paint and brushes.

The "Venus de Milo" (Figure 9.1), even with its broken arms, is considered a major piece of art that speaks unquestionably to the exceptional craftsmanship of the Greek sculptor but also for the wonderful grace evoked in the shown attitude.

---

[1] Some say that medicine is also an art; that could have been true in the past but now we can say it is science and so it has become rational and consequently relevant to the computer.

[2] Or, in such a case, his work should have been quite different.

**Figure 9.1.** Even with its broken arms this statue remains a masterpiece.

Images are an essential vehicle for ritual practices or religious ceremonies because of their fabulous psychological impact in the representation of the world. Some say that they "speak to the soul"! They support for all eternity the power of a virtual reality which could be pleasant or as well, sometimes, frightening.

However, art, even when it gives an impression of mysterious beauty, is not so easily understood in both its technics and in the very origin of the inspiration. As shown in Figure 9.2, the imagination does not reach a conclusion in front of such silhouettes emerging from the past. This piece of art is fabulous from a technical point of view; there is still a deep mystery on how, in those far-off times, the artists have succeeded in carving and moving such heavy stones! Also the aspect of the statues is far from a regular picture of human faces. Was that a kind of "extrapolated" beauty as we can find currently in some modern arts?

The result is quite astonishing, even disturbing and the resulting beauty remains a matter of individual appreciation.

**Figure 9.2.**   The Moai statues of the Easter Islands.

## *Where Does Inspiration Come From?*

This is the most mysterious domain of the brain. There is not any logical foundation for the emergence of the artistic inspiration which surges from the deep subconscious and whatever the kind of art concerned, music as well as painting or architecture. The emergence is purely spontaneous; it may result from accumulated sensations and corresponding appreciation attached; there is not any previous thinking but pure instinct coming from the deep of the mind.

The inspiration is purely personal and remains unique to each individual; this, in some way, is a footprint of each mind and implies to be shared to be successful. Then the artist has to find a kind of resonance with the other people if he wants to be "understood". Sometimes (or even frequently) this understanding is not so easily shared and needs time to get accepted and the work to be admired and valued.

That means that the "inspiration", coming from the deep "soul", translates hidden trends of the mind and so can be considered at the limit of some "abnormality" (as long as this word has a clear definition). Were

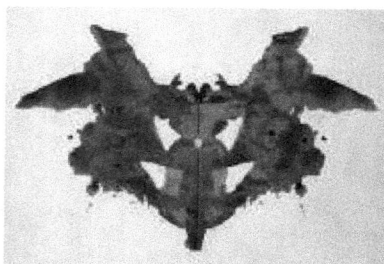

**Figure 9.3.**   A typical Rorschach image.

Picasso or Dali "normal" individuals? Is it normal to appreciate their painting? Was Van Gogh truly insane?

Would the computer have to be introduced in the meaning of madness and would it be doomed to become a little insane to get closer to a human brain?

This trend to analyze the art works has lead psychiatrists to use this way to investigate the bottom of the mind with dedicated tests such as the famous Rorschach test[3] supposed to give a psycho-diagnostic about the personality (namely schizophrenia). As indicated Figure 9.3 an image is obtained from a drop of ink which is pressed in a folded sheet of paper.

The observer is asked to tell what this picture reminds him off. Following the answer the psychiatrist makes the diagnostic of the mental state of the patient! That is the way art could be used in medicine!

## About the Art of Painting

Why and how do humans seek to make Art? At the very beginning "Homo Faber" was striving to make useful objects to help him survive in his very harsh situation. But how did he overcome this basic need to make the effort to make these objects beautiful? A million and a half years before, during the Acheulean stage they produced "handaxes" that reveal an innate sense of beauty through their shape and careful carving. Up until

---

[3] "Rorschach Inkblot Test", Jane Framingham, PsychCentral, October 13, 2018. Available at https://psychcentral.com/lib/rorschach-inkblot-test/

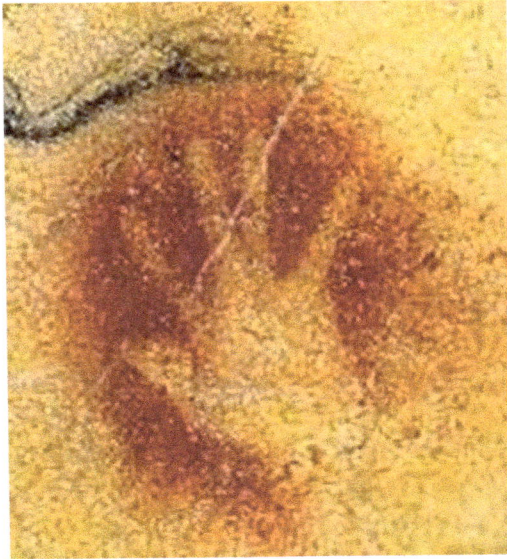

**Figure 9.4.** A print of a Cro-Magnon's hand in the Chauvet's cave, some 37,000 years ago.

today there always will be a competition between the "useful" and the "beautiful" which usually proves more difficult to get.

A very long time ago, an innovative man of Cro-Magnon created a print of his hand (Figure 9.4) by sputtering some colored earths on the wall of a cave, simply with his mouth.

This corresponding picture belongs to the famous Chauvet's cavern. It was convincingly dated some 37,000 years before present. At that time of the Aurignacian stage of the Upper Paleolithic period, glaciation and "global warming" followed. Homo Sapiens survived (they were used to!) and were referred to as Cro-Magnon. It is supposed that this hand is among the oldest of the many pictures of the cave displaying beautiful drawings of animals. The cave was used as a shelter (bears remains) but also as a sanctuary. The entrance of the cave collapsed and the paintings were preserved all along millennia until their recent discovery (1994).

There was no direct need for that negative paint of a hand, just a kind of pleasant artistic picture. So why did Cro-Magnon do that? It could be among the first paintings for humanity, the first attempt to create an image of a natural "object" that will cross millennia. This is pure primitive Art. A very beginning!

Was this a message to the future humanity, saying "Hey Joe! This is my hand, you know!" and nowadays Joe answers: "Hi Bill! I do have found your painting! This was your hand I guess! The same as mine! Good Lord!". More simply it could constitute an address to the contemporaries saying "look at what I have been able to do! You never have seen that!".

Maybe the artist already looked for the admiration of the masses. But at that time they were so far from a competition with the computer! This hand was the first attempt to get an exact representation of the living world; this was a first step of man out of his native bestiality. I guess at that time this unexpected picture could have been somewhat frightening for the visitors of the cavern and considered as a supernatural manifestation! They might have shouted: "Look! The shadow of a hand on the wall! That is magic! Call the shaman to see that!". All things considered, this is as astonishing, meaningful, and precursory as was the first step of Armstrong on the lunar ground; we have also been tempted, in the moment, to shout "this is incredible".

It could be imagined that this surrealistic picture was able to trouble all the tribe. They emerged, for the first time, from the natural world to access into a virtual one, in the deepness of a cave only lighted by the flickering light of the torches. For the first time, a man has created an image of the living as God created it. In spite of its hardiness this image is beautiful by the intuitive significance it displays.

The psychological impact was so disruptive that the shamans of the time used it as a magic sign to convince their contemporaries about the hidden wills of the spirits. Later on, every religion followed the same intellectual path and provided us with holy pictures to be venerated. They even suggested that these statues or paintings possess supernatural powers able to influence the destinies. The Muslim religion was so afraid of the potential power of these virtual realities which could surpass the religious ones, that it strongly prohibited them; they explained that the world Allah created is not to be snapped in a picture by any miscreant. This constitutes there a serious sin to be condemned.

However this primitive shadow of a hand has been successfully followed by more detailed pictures carefully drawn on the walls of the caverns with coal pieces to make detailed representations of animals or hunting scenes. These beautiful cave paintings crossed the millennia to reach us. These naïve and clumsy pictures were abundantly and spontaneously created, independently, in every continent, each one with its own style in order to please different local people.

Then pictures became a social link, a way to exorcise the dangers, a community message to the world, a pedagogical memory that can be referred to. Thus the image has to be as realistic as possible in order to convince people and for the artist to become respected and even admired for his talent to make images pleasant.

## The Consequences

After these prehistorical maturations, came the precursory Mesopotamian civilizations (4,000 years BC) that stem from the basic requirements of the good trades already settled between Asia and Europe.

The cuneiform scripture (Figure 9.5) was invented in order to help keep a trace of the exchanges, with conventional image codes punched in clay tablets. This was not especially a quest for the beautiful but rather a pragmatic technical requirement. That was, anyway, pure invention created from scratch, not any rational deduction a computer could reach.

**Figure 9.5.** The cuneiform first species of scripture.

This was the starting point for a clever evolution toward more sophisticated scriptures that, at the end, makes it able to faithfully follow the spoken. Greek and then Roman alphabets made it possible to translate in a carving or on a parchment any constructed phrase and Socrates later will say: "this is a kind of viewable speaking". Scripture appeared as a learned conventional process whereas speaking is pure instinct through imitation and association of ideas. That was how knowledge begun to be stored, before any computer or Data Center were available!

This was pure technicality but scribes were thirsty for beauty and when we came to the Middle Ages they invented a new art: the calligraphy. The technology has been improved with the use of adapted paints and brushes and the "polychromic" exploded with new blue and green pigments. That will benefit to the religious as well as profane pictorial arts. Printing was not already discovered and the books were hand made (Figure 9.6). These artists were essentially concerned with the perfection of the beautiful.

Painting, strictly speaking, could then flourish with talented artists, from Giotto to Leonardo da Vinci and his enigmatic Mona Lisa, in year 1500. Then came the time of Renaissance with new schools of painting

**Figure 9.6.** The "incunables" were hand made.

from Italy or Holland. Inspiration seeks new compositions and expression in the painting. The style however remains ever exclusively figurative in the various portraits and landscapes displayed. Brain was fully connected and makes sparks!

A flowering of gifted painters widely deploys along these centuries from Rembrandt, Velasquez, Rubens, Watteau, Boucher, and so many others. Then we reached an epoch close to modernity where a new technical invention made a discrete entrance on the stage: the photography which is pure scientific technics set apart the art of choosing the scene for the snapshot. The revolution soon came in the field of painting because photography was more reliably figurative compared to any painting. Then the photography was considered a faithful testimony of the reality which got evidential value.[4] The machine became a very real challenger, so artists had to find another way of processing that the machine could not access.

Then the painters left the traditional classicism and tried to evolve towards other ways of doing; they left to the machine the figurative aspect that was prevailing for centuries and painting became purely abstract or even subliminal (symbolism, cubism, impressionism, etc...); the modern art was born and influenced not only painting but also other kinds of art such as theater, poetry, or novel. A way of doing that photography can hardly compete with.

To make things worse, these modern painters even tried to explore the field of the "ugly" in order to see if there could not be some hidden beauty in this. As the Goncourt brothers, who were art experts, put it: "The ugly, always the ugly, without the beauty of the ugly". These modern painters now congregate in the Bauhaus of Walter Gropius at Weimar (Germany) to show their works; they also claim for an uncluttered style, what Ludwig Mies call by a formula "Weiniger is mehr" or "less is more".[5]

There came the "generative work" which is a trial to collaborate with a computer! Then the machine replaced the hand of the painter. It became possible to make art with Mandelbrot's fractals and other kinds of

---

[4]Of course things have deeply changed with the advent of the computer which now allows creating any fake proof with easy synthetic images.

[5]This is especially clear in the works of Pierre Soulages who essentially displays all the nuances of the color black!

"dynamic art" which can be displayed on screens or printed, no need for canvass. The brain, for the first time, entered in a competition with a machine and that will not be going to stop soon.

However, the competition as about to heat up. The images were allowed to follow a movement and the movies were born. This was to still more enhance their psychological impact.

When the first animated film of the Lumière brothers was exhibited in the year 1896, it displayed a train entering the La Ciotat railway station (Figure 9.7). The realism of this innovation was such that people in the room were really afraid that they could be crushed by this impressive machine and, all the sudden, they quickly rushed out the room.

Then brains had to get accustomed with this new imaging process imitating life even in the movement. Now we are familiar with that, thanks to a special education of the brain. This was a precursory virtual reality we had to get familiar with!

Now television has entered the home and on the computers or the smartphones we are submerged with moving colored images in order to be conditioned to follow instructions, advertisements, or watchwords difficult to avoid. Still remains the nostalgic feeling of beauty.

**Figure 9.7.**   The first film of the Lumière brothers showed a surging train.

## About Deep Understanding

The notion of beauty remains instinctively associated with every machine we are to fabricate. Where could instinct come from to be able to induce intuition, ideas, thoughts, and inventiveness? Rationality cannot do everything; it comes after to give coherence to the thoughts. We don't know how the brain, throughout all these millenaries, has evolved, on its own, to get a structured knowledge (we call it education) and generate a shared culture? Was it a genetical slow mutation? That assuredly has been the result of a collective effort where the sense of beauty had to play a noticeable role to make the best choices, even in the technical domain.

Would the computer, one day, do the same and follow a similar track to reach such a "personality"? Would computer be able to generate spontaneous "genetic" mutations? Without any doubt that will not require millenaries of maturation!

The first "planes" of Clément Ader (Figure 9.8) or of the Wright brothers were especially ugly, could that be the reason why they had so much trouble to take off?

The technical know-how was still in infancy; now any engineer in aeronautics would say that if a plane is not beautiful it will assuredly not fly well. If the machine wants to equal man it would obviously have to take that into account and become intuitive and sensitive to beauty.

To come back to painting, the artists of the 2.0 world have completed an exhaustive review of every possibility and they finally turned to what

**Figure 9.8.** Eole, the first "avion" of Clément Ader.

**Figure 9.9.**    The street art of the 2.0 world.

is called the "Street Art" which displays a full creative freedom on a wall in a street (Figure 9.9).

So, looping the loop, after so many millenaries of painting, we have rediscovered an art which is nothing more than the cave art of the Cro-Magnons. The technic is the same except that the artists use a paint bomb instead of sputtering the paint with the mouth! Would that be the best method to achieve beauty? The leading difference remains that the Cro-Magnon's work was strongly figurative to make an image of a hand whereas Street Art's painters represented just anything at all. In any case we could be sure that we would have experienced every solution.

### Now, What is on the Road?

What is to be expected from the computer in its challenge with the brain? In our 2.0 world, human craftsmanship is on the way to be replaced by a machine technology that began to be more and more "intelligent". Would that be enough to compete with the human creativity? I would answer "currently not" but things are changing so fast!

Currently software means have obtained an impressive level of "sophisticatedly" modifying the images and the sound in real time to the point that seeing a personage on the screen, talking in an assembly, does not guaranties that this event really took place or is pure Virtual Reality. No doubt that could be pure fiction. Even the facial movements and the personal accent of the speaker can be mimicked flawlessly in quasi real time. The Artificial Reality in the images is so perfect than it is called Deepfake. But, nevertheless, the leading idea never comes from the computer. This is an open door to hoaxes and especially "infox". However this is in no way a proper initiative of the computer which has to be taught to do so.

These technics could be considered as an Art, to some extent; it has been used and did lead to abuses[6] in various domains from politics, marketing, or even to pornography! A time is to come when human observation will no longer be able to determine what is true from what is fake.[7]

## The Painting Computer

The computer already took up the challenge to become a pure artist in painting[8] with the visual art processing. Closely related to the progress that has been made in image recognition is the increasing application of deep learning technics to various visual art tasks. DNNs have proven themselves capable, for example, of

a)  Identifying the style period of a given painting,
b)  Neural Style Transfer — capturing the style of a given artwork and applying it in a visually pleasing manner to an arbitrary photograph or video, and
c)  Generating striking imagery based on random visual input fields.[9]

---

[6] "Terrifying high-tech porn: Creepy 'deepfake' videos are on the rise", John Brandon, *FoxNews*, February 16, 2018.
[7] "Wenn Merkel plötzlich Trumps Gesicht trägt: die gefährliche Manipulation von Bildern und Videos", Tomislav Bezmalinovic, Aargauer Zeitung, February 3, 2018.
[8] A conference was hold in Atlanta in June 2017: The 8th International Conference on Computational Creativity.
[9] "The Machine as Artist: An Introduction", Glenn W. Smith and Frederic Fol Leymarie, Arts, 6(2), 5, 2017. "Art in the Age of Machine Intelligence", Blaise Agüera y Arcas, *Arts*, 6(4), 18, 2017.

**Figure 9.10.**   An unknown personage born in a computer.

In 2016 Microsoft researchers associated with the Dutch university of Delft succeeded to create a purely invented portrait of a man as Rembrandt could have done in his time (Figure 9.10). The computer studied in detail the way Rembrandt was painting, after analyzing with its Artificial Intelligence some 346 works of the Dutch Master.

Then its conceptual unit imagined a virtual and unknown personage in the Rembrandt style and the painting was processed by a 3D printer and special paints, in order to retrieve precisely the brush stroke of the real painter in the thickness of the paint. The personage in itself was not copied in some archive but purely invented by the computer following the imaginative general aspect of a possible Dutch man of the time. In some way we could say that the computer was capable of some creativity even if it was guided in its choices. We are assuredly here at the edge of an imaginative work close to a very brain production.

The result was so stunning that qualified experts were ready to accept it as an authentic Rembrandt painting emerging from the past. This canvas is in no way to be considered as a copy or a fake; rather a pastiche, an accurate imitation or a pure invention of what Rembrandt could have been doing centuries ago. The computer really slipped into the skin of

**Figure 9.11.** The artist is now an intelligent robot named Ai-Da.

Rembrandt to reinvent his painting of an imaginary personage. The artist has gotten a second life through a machine. This is "beauty" in the Rembrandt's way which resurrects unaffected after 300 years! That does not mean that the machine has become a living new Rembrandt it still remains in some way "a follower".

One may now foresee that, one day ahead, a computer looking for the beautiful should release his own works that could not be inspired by anybody. That is quite chilling but, you know, we are in a similar panic to the spectators of La Ciotat frightened by the train in the movie! Algorithms based on "deep learning" lead the computer to make its own "artist culture" ahead of a notion of beuaty yet to be defined.

However this news has been recently outdated by a robot humanoid artist that succeeded in making its own painting with its artificial mind and hands (Figure 9.11). She (was it a female?) was named Ai-Da and comes from a laboratory in Cornwall UK (and not in Japan however!).

I wonder what she (it?) intended to paint but it comes from its own inspiration and realization. It's up to her (still it?) to decide what to represent on the canvas.

No matter what a man or another robot could appreciate in these "artificial" and unique paintings it still remains that they had a good sale, for more than a million Euros, at the Oxford's Barn Gallery.[10] After analyzing some 15,000 classical works, the "brain" of this artist executes an algorithm simulating the human brain through a couple of neural networks required to be at war with each other. These works are unique. This belongs to the "generative" art which intends to extend to fashion, music,[11] or even literature. Where would we go too far?

## What is the "Predictive" Text?

Painting is not the only domain where Artificial Intelligence is to compete with the human brain. We have entered in an epoch where the work in AI could be compared to medieval practices of the science wizards. Are we to await the arrival of an Enlightment age to make current alchemy vanish and reach a very perfection?

Science is bombarding us with incredible news. New theories arise which foresee the extension of the life, a possible emigration on Mars or a biologically improved man. That is the theory of Transhumanism which currently takes place. The machine is not yet able to get a real "sentiment"; it remains so impalpable that scientists are struggling to teach that to the computer, but that is to come, not in a distant future.

In May 2018 Paul Lambert (at Google) introduced a feature named Smart Compose which suggests to people typing a text on a keyboard what they have to extrapolate afterwards. This software has been swiftly followed by a more sophisticated version: Smart Reply. The AI of the computer previously informed of your personality analyzes your customer wishes through the words you have written and wraps up by itself your sentences as a "predictive text". As John Seabrook[12] said "That's one small step forward for artificial intelligence, but was it also one step backward for my own?". He was really fascinated by the way "the A.I. seemed to know what I was going to write." or "Sometimes the machine seemed to have a better idea than I did." or even "how long would it be

---

[10]They also sell well from Christies!

[11] Some years ago Ianis Xenakis invented the "stochastic and spatial" music programmed in Fortan language and played by an IBM computer of the times!

[12]"The next word: where will predictive text take us?", John Seabrook, *New Yorker*, October 14, 2019. Available at: https://www.newyorker.com/magazine/2019/10/14/can-a-machine-learn-to-write-for-the-new-yorker

before the machine started thinking for me?". These remarks are in a perfect agreement with the response of a "top gun" pilot challenged by a simulator. He said: "I had the disturbing feeling that the machine was aware in advance of what I intended to do".

Then he suggested that, for the moment the only way to communicate without being spied on by the machine would be to use hashtags or emojis as a modern version of the hieroglyphs! Paul Lambert[13] explained: "At any point in what you're writing, we have a guess about what the next number of words will be". But if the AI is well informed about you (through analyzing all your e-mails, for instance), it could "ramble on longer".

Recently OpenAI, an AI company, announced the delay of its AI writer GPT-2, "because the machine was too good at writing." Three years ago the machine translation was "too error-prone to do much more than approximate the meaning of words in another language" but switching to the neural machine changed the game in 2016 and Google has replaced human translators in many domains and "Google Translate" is found accurate enough to rely on. "Now a piece of writing is no longer a certificate that an actual human is involved!". As a matter of fact compute grows even faster than Moore's law and the only limit of it lies currently in the consumed energy in the "server farms".

In this way it is worth noting that, up to now, in the advances of the civilizations there has always been steps generated by the discovery of a new source of energy; this is true from the coal to make fire to the advent of the nuclear energy to largely provide electricity. However the new 2.0 world has been in no way engendered by a new kind of energy which remains to be discovered to help Data Centers survive.

## What about the Vision of the Philosophers and Forecasters?

Transhumanists are the philosophers of a realistic future, as they claim are not being rooted in a Science Fiction framework. Julian Huxley[14] (a brother of Aldous) was the first to use the word "Transhuman" (in 1957)

---

[13] From Smart Compose at Google.

[14] "Julian Huxley, Le Transhumanisme, 1957", see: http://sniadecki.wordpress.com/2015/01/21/huxley-transhumanisme/

to describe "a man who remains a man but transcends himself by deploying new possibles to and for his human nature".

Transhumanists try to have a realistic judgment on the biology and independently on the computer which both are in a phase of fast implementation. They belong to very different nature, then how to meet them together? Meanwhile "posthumanists" bluntly emphasize that downloading a spirit in a computer would likely be feasible in a near future thus making the machine quite indistinguishable from a human brain. The question is not clearly clarified how they imagine the final status: would the machine become independently fully autonomous by itself or would it require a mixing with a human brain "improved"? Thus they assume that intuition, arts, creativity and the like might be simulated in dedicated software.

Artificial Intelligence has become a strong competitor challenging the human one previously considered as definitely superior. This field is a major one in the transhumanist philosophy. But remember that finally, in a prophetic vision, Darwin said, long ago, that: "it becomes probable that humanity, as we know it now, would not still reached its final state of evolution but more likely a phase of beginning".[15] This introduces the next paragraph!

## Would Current AI Still be in a Medieval State?

As we said, the brain is estimated to contain a hundred billion neurons, with trillions of connections between them. The neural net that the full version of GPT-2 runs on has about one and a half billion connections, or "parameters". At the current rate at which compute is growing, neural nets could equal the brain's raw processing capacity in five years. To help OpenAI get there first, Microsoft announced in July that it was investing a billion dollars in the company.

However, Ilya Sutskever[16] added: "Researchers can't disallow the possibility that we will reach understanding when the neural net gets as

---

[15] "A history of transhumanist thought", Nick Boström, *Journal of Evolution and Technology*, 14(1), 1–25, 2005.

[16] Currently Chief scientist of OpenAI, computer scientist working in machine learning.

big as the brain." and Frank Chen[17] announced that "this is deep learning Cambrian" to mean that we are still far from the coming of a new age for AI. But Nilsson[18] relativizes these successes: "AI has by now succeeded in doing essentially everything that requires 'thinking' but has failed to do most of what people and animals do 'without thinking' — that somehow is much larger".

This raises the question of the consciousness of the computer and its possible ability to stay in relation or in dependence (not to say in symbiosis) with a human brain. Machines are organized as rigidly structured hierarchies of logic modules, whereas biology is based on holistically organized elements in which every element affects every other. Even if schematically presented, only biological systems can use this design principle. Also to be mentioned is that, in a computer, the hardware and the logical set-up are fundamentally distinct entities whereas the brain is not a dualistic entity and is more complicated than simple logic gates; this actually makes the brain more difficult to upgrade.[19]

The ability of a human brain to instinctively make significant relations between concepts which could lead to confusion is presently hardly challenged by machines which still are dominated by a stiff logic even concerning simple events such as those rising in a trivial sentence. For instance, the famous sentence "spirit is ready but flesh is weak" could easily be "understood" by a machine as "the vinegar is served on the table but the meat is of poor quality" if reference is made to a cooking situation. What we call "the common sense" cannot be so easily modelled and translated in an appropriate algorithm even if our brain is familiar with, in an instinctive natural process. There is assuredly no hope to hold out that in a near (or remote) future a computer will be able to provide us with a significant translation and interpretation of a quatrain from Nostradamus!

There is no doubt that pattern recognition, facial expression, or language analyses are currently in a booming growth. Would that be enough for these deterministic methods to compete with the "plasticity" of our

---

[17] Frank Chen, Partner at Andreessen Horowitz — Machine Learning and Deep Learning.

[18] *The Quest for Artificial Intelligence*, Nils J. Nilsson, Cambridge University Press, 2009.

[19] "Levels and loops: the future of AI and neuroscience", Anthony J. Bell, *Philosophical Transactions of the Royal Society of London B*, 354(1392), 1999.

spirit? Would it be mandatory to teach the machine the notion of good or bad, pain and death also with a dose of fear, a dose of moral, somewhat similar to the robot's limitations and warnings previously requested by Alan Turing? Intuition is not so easily able to be "modeled". As Voltaire put it, "Man invented God, not the opposite".

Religions fall within the area of the imaginary and creativeness of the spirit. Similar to the Arts they do not belong to any logical mechanism. How to teach that to a "mineral" machine? Neurophysiologists carefully study the way to have an open access to the unconscious but without reaching any definitive answer. They emphasize an external interaction with the brain and recognized that sounds and notably infra-sounds play a key role to the access of second state of mediation. Sleeping and dreams also belong to the remote and unknown activity of the neurons in the brain. That's all we can say for the moment.

## One are Concerned with Man but the Issue is Men

Humanity is in no case a collection of identical human beings. They strongly differ from a continent to another one and inside a continent and even, at a still smaller scale, inside a given society or family (even between twins). Each one enjoys his own personality which is quite unique. What is established to be appropriate for one will not necessarily suit the other. This point is rarely addressed by transhumanists; how to get it? The mind is so fugitive, so plastic, so changing, that putting it in a holistic description remains illusory and a fortiori to translate it in a machine language. There is not any absolute reality of what a particular mind (at a particular instant t) could be summarized in a snapshot.

The main concern with men is "intelligence" which is, as a matter of fact, considered as making the difference with the Artificial Intelligence attributed to computers. There have been many efforts to quantify the human intelligence, so diversified! But the result remains fuzzy; how to evaluate a parameter we are not able to give an accurate definition of?

The most familiar test is the Intelligence Quotient (IQ) which consists of a set of questions which are supposed to deal with the whole issue of the culture of the candidate. But unfortunately that is a simple question-naire which in no means be able to reveal a potential intuition or creativity (especially when dealing with the domain of the arts). Thus an IQ test is perfectly appropriate for a computer but irrelevant to properly quantify a real human intelligence level. IQ is poorly adapted to check an artistic predisposition.

In the prospect, dear to the transhumanists, of a biologically "enhanced" man would this "enhancement" be convenient to everybody or will it be restricted to selected people, the other being considered as "chimpanzees" of poor IQ, as someone suggested?[20] Would this enhancement imply a permanent link with a computer as it already happens with the current invasion of the smartphones? Would reciprocally an intelligent computer require as well a permanent connection with a brain in order to follow what it is not able to find by itself?

To conclude this showdown on the "man *versus* the men" it may be said again that being super-intelligent does not mean being super-powerful; "the speed and the viscosity of the rest of the world will limit physical and organizational changes ... physical changes will take time" (Max More).[21]

---

[20] *Remaking Eden: How Genetic Engineering and Cloning will Transform the American Family*, Lee M. Silver, Ecco, 2007.

[21] "A critical discussion of Vinge's Singularity concept", David Brin *et al.*, in The *Transhumanist Reader*, Max More and Natasha Vita-More, eds., John Wiley & Sons, 2013.

# Chapter 10

# How to Make Man and Computer Cooperate

Increasingly, man is undoubtedly to live with more or less advanced, possibly humanlike robots of various forms. The skills of these machines will grow inexorably, to the point of challenging the human.

Many efforts are undertaken in different places to get a better understanding of the brain and its simulation in a machine, but these efforts remain partial and scattered; a global solution cannot yet be emphasized. How does this man–machine cohabitation look today? Would man be able to stay unchanged? Would his biological nature have to mix with mineral components? Would he have to mute biologically? Would these changes be compulsory for all or reserved for a selected population?

Would man simply be replaced by machines? Here are the questions now opened up.

## How Do We Take Advantage of the Computer?

### Bio–Mineral Interfaces

The key prerequisite for communicating intimately with a computer lies in the bio–mineral interface: two different worlds are brought together, each with its own specifications which will not be easily matched. So the successes stay limited and specific (at the moment).

Let us consider some recent examples of such implanted devices, keeping in mind that there will in any case remain the issue of the sustainability of these mineral components subjected to the constant rejection response of the body.

- In 2012, Jan Scheuermann, 55, a tetraplegic woman, was implanted with two electrodes in the brain, which made it possible for her to feed and drink by herself. More recently, she was even able to command by thought an F35 simulator using virtual commands (DARPA)!

- Deep brain stimulation was experimentally used to cure Parkinson's disease or to try to identify and connect the very source of the brain waves. Also, implants work in the treatment of strokes and traumatic brain injuries. They take signals from one part of the brain, process them conveniently, and then stimulate another part of the brain in order to bypass the damaged zone.

- As far as Parkinson's disease is concerned, the electrical compensation is normally devoted to major cases. Electrodes are implanted deeply in the motor area (hypothalamus) on each side of the two hemispheres, as for a pacemaker. Satisfying results are obtained but they remain at the experimental level; secondary effects or drifts often occur. The operation requires image processing assistance and, to be more accurately guided, should be performed by a specialized robot.

- The hippocampus, for its part, manages the memory; it plays a crucial role in the aging process and also in Alzheimer's disease. Experiments have been carried out successfully by DARPA (Defense Advanced Research Projects Agency) and USC (University of Southern California) with monkeys and mice to stimulate the failing activity of the hippocampus. It is expected that artificial hippocampus devices could be attained, thus providing a "hard drive" equivalent of the brain, similar to a "backup" for the memory. Theo Berger (USC LA) and Joel Davis (Office of Naval Research) are confident of successfully developing implants dedicated to skill enhancements soon.

- Always in the field of deep-brain-implanted electrodes, it is worth mentioning an experiment successfully performed by Miguel Nicolelis (Duke University), with a macaque, to remotely command a robot replicating the walking movements of the monkey. The implant was placed in the brain cortex of the macaque (Durham, North Carolina) and the robot was located in Japan!

This opens the way to extrapolating the experiment on paraplegic humans in order to allow them to command, for instance, an exoskeleton or even an artificial aide (android). The experiment could be extended to moving a mechanical arm or, more tricky, to "fingertesting" an object, even a virtual one. The researchers had previously emphasized presenting a couple of tetraplegics equipped with exoskeletons to kick-start the Soccer World Championship in 2014 but, unfortunately, they had to ditch that at the last moment.

- Similarly, Neurobridge has been proposed for paraplegics: a matrix of 96 electrodes is installed on a 3.8 mm chip which is implanted at the right place in the brain to selectively collect the basic neural information corresponding to the movement of an arm (for instance). The signal is then transmitted to a laptop, which conveys the amplified signal to a "sleeve" of electrodes on the forearm, which stimulate the corresponding muscles and activate the hand. Thus, the useful signal can then bypass the wounded region of the spinal cord.

Of course, it remains quite difficult to decipher the way the brain coordinates so accurately a movement which sometimes could involve a lot of different muscles and a corresponding sequence of orders to master a three-dimensional complex displacement. The foreseen action has to be quickly and accurately planned before being launched. We are still expecting to exactly reproduce the mechanism in a computer to drive corresponding actuators. This has begun to be done with a monkey but not yet with a human.

- More recently, DARPA reported[1] a successful experiment to develop an artificial hand which conveniently restored the sensation of touch. A paralyzed person was equipped with electrodes in the motor cortex, which controls the thinking, as well as in the sensorial cortex, a region which is related to the sensation of touch. The patient was perfectly able to control his fingers and to appreciate the contact with various objects, identifying exactly which finger was involved.

- Among the various brain implants, the retina ones[2] should also be mentioned, because the retina is part of the brain. A matrix of electrodes is implanted in the eye in order to connect a "neutrophic"

---

[1] 01net, No. 828 (September 2015).
[2] Amine Meslem, 01net, No. 819 (2015).

external camera with the optical nerve. This way the retina is bypassed, thus providing compensation to age-related macular degeneration. The camera acts as an intelligent interface which adapts the image to be transmitted in order for it to be accepted and processed by the brain as a natural image. Of course, instead of trying to implant an electrical link (or an optical fiber) in the brain, an easier way would be to capture the brain waves externally with the help of adapted antennas directly placed in a helmet on the skull; the inconvenience of this method would be that, maybe, the signal could be polluted by other sources located inside the brain. In such a mess, discriminating what is what and what comes from what requires a delicate signal processing.

## The Thought-Driven External Command

This leads us to the thought-driven external command. Some examples are given below, with discussions:

- BodyWave is a simple gadget, a demonstration of which was given by Greg Gage at a TED[3] talk. A brain nervous signal is collected through a helmet and wirelessly transmitted to the muscles of another man, thus substituting the command for the will of the recipient.

  Could this method be extended to transmitting signals other than a muscular command, and so share or even impose unwanted ideas or wills through a second helmet on the skull of the recipient? Such a possible transmission of thought is not really obvious, but it is not to be sneezed at. Such mind-reading devices interfaced between EEG sensors and a computer or another brain[4] would provide a means for mind-to-mind transmission.

- In September 2015, a paraplegic[5] who had severed his spinal cord was equipped with such a helmet and trained to stand up alone, without the

---

[3] Technology, Entertainment, Design.

[4] Such a brain–machine interface (BMI) was proposed by the Japanese, likely to help the deaf and dumb.

[5] "Person with paraplegia uses a brain-computer interface to regain overground walking", Youtube, April 21, 2014. Available at: www.youtube.com/watch?v=RBAeR-Z0EHg

use of robotics; that worked and it is expected to be able to extend the experiment to making him walk freely. "Stand up and walk," said the Lord to Lazarus — the miracle is almost to be trivialized! It was said that in the first step the recorded waves corresponding to the command were produced following a voluntary concentration order from the brain while the second step would be to train the machine as well as the man to make the subconscious participate in the operation as it does in natural walking. The I/O bottleneck of the operation[6] comes from the requirement to translate the brain's alpha waves into a digital language the computer can understand.

All of these possibilities of "externalizing" brain waves and inducing back-and-forth cooperation between the brain and a computer also would drastically shorten the delay in educating or training a brain — but are we still in science fiction? To go a step further, whole brain emulation (WBE) was proposed in the 1950s to make a digital copy of the brain in a computer (Ray Kurzweil is very fond of this idea). Actually, it would be a highly adapted solution to help with space travel. Would it also be a solution (also following Kurzweil) for the immortality of a "digital you" (DY)? We will deal with this later.

In creating a DY, many unexpected and somewhat incongruous questions will arise. We briefly and non-exhaustively suggest some:

- Would destroying a DY be regarded as murder?
- What kind of physical support is to be preferred — humanlike, electronic box, or even purely virtual?
- What about copies, legal properties, patents, and so on?
- What to do with the DY if the original dies?

As long as we do not know more about the brain's organization, there is not any hope to fully replicate something we are still ignorant of.

However, might I introduce a personal view? Considering that my computer is already an appendix to my brain, I am eager to get a friendlier interface!

---

[6] Gerwin Shalk, MIT's Emerging Technology Conference (September 2015).

## The Internet and the Cloud

Of course, following the extraordinary accumulated resources of knowledge currently accessible on the Web and the cloud, it becomes very tempting to imagine bridging the gap with the brain. The many advantages of the computer cannot be contested: speed of calculation, incredible memory access, and worldwide transmission capacity, all of that without the inherent complications of the biological world: infections, intolerances, bleeding, or irreversible brain damage.

At the moment, we have to make do with conventional access through a keyboard or at best vocal commands, but things are likely to change in a not-so-remote future as long as thought-driven communication technics progress.

## Progresses in Brain Surgery

Biological access to the brain, as we have seen earlier, is not so easy. Implants suffer drastic difficulties in cohabiting, at least in the long term, with biological tissues. In any case, they irremediably become obsolete very quickly.

Cerebral grafts or transplants, for their part, are rarely accepted and are non-necessarily adapted for the purpose. Brain surgery obviously is continually improving, following better knowledge of the brain, but is still close to the experimental domain. Needless to say each brain has its own distributed organization, which differs in the details from another brain, and will hamper efforts to efficiently intervene surgically.

All of that is to say that the ultimate goals, such as therapeutical regeneration or brain reconditioning, are still in their infancy and that "whole brain prosthesis" is still in the domain of science fiction. The main interest remains more focused on the cure for brain disease, which is still often a mystery.

Brain-to-brain communication would be possible the day when the brain could reciprocally be receptive to electromagnetic waves. Would language remain the only solution? We do know that it works fine without any surgery!

Entering a digital action into the brain through electromagnetic waves remains to be discovered even if, in the opposite way, we slowly begin to understand what the waves emitted by the brain mean. Such a lack of knowledge, however, has not prevented business from flourishing. An incredible market has arisen[7] with the "brain training" devices which promise a 10% improvement in "brain fitness" to the average Joe. These much-diversified wearable gadgets control the brain's activity through EEG and can also provide electrical pulse stimulation to make you calmer or more focused.[8] Amy Web (Webbmedia Group Digital Strategy) commented, "You might adjust your mood as the temperature of your living room"!

Of course, to date, nobody knows exactly where to put the electrodes, or how to adjust the level of these "transcranial direct current stimulations." Rex Jung, a neuropsychologist at the University of New Mexico, put it thus: "The science is not there to warrant efficacy." In spite of all these uncertainties and amid the ignorance of possible side effects, the market has still grown to US$1.3 billion!

However, even though progress has been steady in brain surgery, it remains that when one is putting fingers in the brain — to remove a tumor, for instance — a major difficulty is to discriminate what can be removed from what must be preserved, and this boundary often appears quite empirical. A recent technic relies on the involvement of the patient himself, who is kept awake and very conscious. He is equipped with a virtual reality headset which enables the surgeon to test the patient's mental awareness in concert with the surgical act. This allows one to precisely locate sensitive areas.

## How Computer could Benefit from an Improved Knowledge of the Brain?

The first consequence of a better knowledge of our brain is obviously to stimulate the achievement of a computer copy of the brain's organization through efficient neural networks. From that point on, they could provide an enhanced intelligence, if not a super one, simply by organizing and improving access to knowledge (Internet).

---

[7] "Electrify your mind", Jennifer Alsever, *Fortune*, December 1, 2015.
[8] Some names: Melon, Emotive Insight, Melomind, Narebis.

As mentioned before, a key issue also remains in the understanding of the way the brain manages the coordination of the movements. This knowledge of the neural networks would immediately result in impressive progress in robotics.

# Changes are Already Underway

From all of that comes the fact that computer(s) will play an increasingly important role in our lives, and infringe on our liberties and behaviors, and that raises a lot of questions. Would we have to accept the future, and is there a choice? Beyond a technical balance, would we be allowed to ask ourselves some questions about our future already engaged? Would we be placed on an uncontrolled collective toboggan which would change our lives; like it or not?

## *Computers, Phones and the Like*

Strange habits have seeped into the education of kids; for instance, they use touch tablets or PCs in place of paper more and more, and thus they forget how to use their fingers to write or to draw a picture. Do we really regard this as progress? Of course, it is important for kids to get accustomed to the computer in its up-to-date version, but is there any reason not to be educated in using their fingers? Or will the fingers be of no use any longer in a future life? Does writing or hand drawing have to be classified as belonging to the prehistoric ages?

In the same way, the use of fingers seems to be restricted to touching the screen of a smartphone in anticipation of a fully voice-controlled phone. I must admit that kids have attained fantastic dexterity in entering their countless messages!

In general, we all are dependent on our inseparable and ubiquitous phone, to the point where we could raise the following question: Mobile phones are no more than 10–15 years old — how did we previously live without them? It is really incredible how, in so short a period of time, this object has become so essential for everybody on earth, overturning all the individual as well as collective behaviors!

Anyway, the phone can now be provided with a multitude of so-called "applications" (generally free) which aim at resolving any issue of daily life. Each company has its own gadget: Apple (Siri), Google (Now), Amazon (Alexa, which comes with a "Totem"), Microsoft (Cortana), Facebook ("M"), and so on. These "personal assistants" give you train or plane timetables, make reservations, remind you of the shopping list, and so forth.

However, at the same time, these friendly software programs, in the aim of providing you with better, and closer service, behave as a Trojan horse filled with AI soldiers ready to absorb every piece information, even private ones; from this data mine they can deduce your religion, your health, your income, your sex.... Everything which is written in a document or a piece of mail, everything which is said in a phone call, every name or address among the contacts, every piece of valuable information is targeted, classified, and carefully stored. Even your GPS faithfully discloses where and when you went, and also often what for. This combined operation turns out to be the worst individual espionage no totalitarian regime would ever have dared to build! On top of that, this is perfectly accepted!

## Google's Curiosity

Among the various threats to our liberty, Google takes the lead because of its extended curiosity and its extensive, sprawling means. The information about individuals is gathered at a fast pace from the various systems Google has implemented around its more and more intelligent search engines, and this is not going to slow down. In a not so remote future, Google will know in detail everything about your life and your personality (habits, preferences, opinions, health, etc.), and it will be very difficult to escape its night-and-day investigations. No need to analyze your brain waves. Formally, this is all for your own good; Google "is not evil," as it was said.

But Google is not alone in the large scale search for personal information; it is worth mentioning the more secret and quite "stealthy" NSA,[9] a government organization which plays a similar role (more discreetly maybe, but just as efficiently). Its large ears are open worldwide and scan the sky of satellites to record every piece of informative talk. How it uses

---

[9] National Security Agency, dubbed "No Such Agency."

(or can use) its information is a strictly classified secret. But Will Ackerly[10] has warned us, "If you are in their sights on an individualized basis, it's 'game over'!"

The issue, however, remains to know to what extent these "mines" of data (including your medical file) are out of the reach of hackers. The recent story of Edward Snowden and the NSA demonstrates that leakages or hacking cannot be prevented. This is a permanent threat depending on such sources of juicy information, and we all are potential victims of ill-intentioned people. What could happen if, one day, "our very brain is in the cloud"?

Ackerly is of the opinion that the only protection against cybercriminals and shielding people's information lies in encryption. He invented a hacker-proof protection called the "Trusted Data Format" (TDF), which, he claims, "makes it dramatically harder for anyone to pilfer what the user wants to protect". This could be especially effective in protecting medical files while allowing an easy exchange of them. This seems so much more efficient that the U.S. government wants to use it. I just wonder for how long it could be a "definitive" answer in this game of cops and robbers!

## Virtual Reality / Dialog with the Objects

Virtual reality also tends to replace the real surrounding world by an artificial preconditioned one, and hence substitute a guided will for our free choice. This too is done initially with the best intentions, but being so nursed is likely to turn us into a ragtag band of idiots and pure, sheep-like followers.

Also, we are threatened, at home, into preferentially talking and communicating with objects rather than with other humans, such as our kids. That is already emerging with the invasive domestic PC, while our kids are immersed in their smartphones. We are on the verge of becoming slaves of machines which will dictate their laws. We in the Western world feel rather reluctant, at the moment, to get familiar with Pepper, the robot, but the Japanese are very empathic to such a buddy. Robots now play an increasing role in human activities: in the factories, in the hospitals, in the planes or the cars, even at home! All the more so as they are on the way

---

[10]"The anti-hacker", Luke O'Brien, *Fortune*, October 1, 2015.

of "being intelligent". This of course generates lots of new social, juridical, or medical problems. I just discovered[11] that a motion was put forward at the EU Parliament in order to take robots as "electronic persons" and define their corresponding rights and duties!

### Is Changing the Brain and the Mind Underway?

This is assuredly the case, and nobody would ask us for our opinion. Changes will certainly arise and be imposed by the collective life, and the media will obviously play their role in managing a "consensus." Brainwashing is underway and the machine will be there to help. No need to transfer our brain to a computer to make the AI machine a despot. Nobody would ask us for our opinion.

However, at the moment, we still prefer to negotiate important agreements with a person (and not a machine), hearing every quaver in his voice, noting when he crosses his arms, looking into his eyes. We also much prefer, when we are sick, to hear the diagnosis from a (human) doctor, perhaps just talk, and know we are heard by a human being. Meg Bear, Vice President of Oracle, said: "Empathy is the critical 21st century skill." However, Watson is here now, a highly qualified doctor-bot!

## Androids Already Exist

This subject will be complemented in the next chapter, but we can already mention it here.

Robots now enter into our daily lives; they can replace man in many selected circumstances, but never in a whole function of a human. We know how to make a machine speak (even reply to a human in a limited context), we know how to make a machine work in a factory (given a specific task), we know how to make a machine play ping-pong (not yet tennis), we know how to make a machine drive a car, we know how to make a machine find information in a mountain of data. These innovations are very recent and also amazing. Robots often carry out their work better than a human could. However, they all are related on a specific and limited

---

[11] http://reinformation.tv/robots-personnes-electroniques-motion-parlement-europeen-dolhein-56949-2/

application; the question now is how to make the machine as versatile, intelligent, and polyvalent as a human can be. This is the old dream of Pygmalion: give life to a material assembly and possibly make it smile (as Mona Lisa)!

This ultimate grail would of course necessitate accessing a brain equivalent, as flexible, subtle, and effective as a human one, and we know that many projects are heading in that direction. However, the brain is devoted not only to thinking, but also to (maybe this is its more important role) managing the relationship (both directions) with the outside world. This is quite a challenge.

The control of the movements and gestures by the brain remains a mysterious process. We do not know how the brain proceeds to give (and control) its orders to the muscles with such precision, even anticipating the movement of the object to catch it, and this command is extremely complex because it is also closely linked to sight, hearing, or touch.

Many attempts have been made to mimic that function; a particular interest is in the touch control (also called "haptics") and tactile feedback. Promising results have already been obtained: the mechanical sensation would also be accompanied by a "sensation" of hot and cold, dry and wet, or oily, and would be especially useful in surgery, for instance; impressive results undoubtedly have already been obtained without being able, to a large extent, to compete with a human finger. A very good advantage of the robot over the human hand is certainly that the bot does not tremble and is capable of keeping a micronic positioning and of reproducing gestures that are exactly the same. Both the biologic and the material world keep their respective advantages but we are still very far from duplicating Mozart in a machine!

Nevertheless, the advancement of the researches cannot be stopped and very impressive results have been obtained in the implementation of humanlike manikins, which more and more closely resemble men: androids.

### About Bots and Androids

At the very beginning of robotics, Alan Turing proposed a simple test to determine if a robot has reached the human level of intelligence.

He imagined a conversation between a human and a robot through a phone line. Would the human be able to discover he was talking with a machine? This criterion could have been fulfilled in a recent experiment involving a set of English-speaking participants and a machine representing a young Ukrainian not exactly familiar with English. The illusion was achieved for a large majority of participants.

However, even if that step is satisfactorily fulfilled, robots must be carefully restricted in their behavior in order for them not to become a danger to humans. This is called "friendly intelligence." In that way, before robotics became a full-fledged science (in 1942), the novelist Isaac Asimov imagined the "three laws of robotics," which shortly afterward Turing adopted for keeping a friendly relationship with humans:

- A robot may not injure a human or, through inaction, allow a human to come to harm.[12]
- A robot must obey the orders given to it by humans, except when such orders would conflict with the first law.
- A robot must protect its own existence as long as such protection does not conflict with the first or second law.

This is quite relevant, but nothing is said about the ways law enforcement would be applied if one of these points was not respected. Laws are meaningless without the corresponding sanctions (and police!). However, a bot ignores what we call pain or death. Would these notions have to be taught too, and would they have to be considered as an indefectible part of intelligence? How do we make a bot feel pain? Or death? How to "sanction" a computer, to make it feel guilty? Pull out the plug?

In the human case, pain is somewhat useful in the sense that it works as a warning against coming dangers. That could also be profitable to a robot during its active "life" — but how to make a robot learn pain? This is, however, partly included in the first Turing law.

Another point is: how do we explain to an honest robot the how and why of making war?! This is a very important point, because one of the

---

[12] That is, "Thou shalt not kill," in the Christian language.

main applications emphasized for the robot (already effective) is to make war in place of humans. How do we conceive a drone's humanlike intelligence to kill enemies and differentiate between friend and foe? Of course, if so, it would be quite humanlike (including the unavoidable human errors)!

Every year, an international competition takes place in Los Angeles under the sponsorship of DARPA, where the best up-to-date bots compete with one another to compare their performances or display new applications. The best robotics labs confront each other: Google, MIT, JPL,[13] NASA, and so on. Also, foreign contributions are welcome — from Japan, South Korea, China, Italy, and Germany.

From year to year, spectacular progress has been achieved but much work remains to be done before reaching a satisfying solution, given the huge variety of situations envisaged. The topics addressed focus on the robot which could be able to perform the same task as a human but in a hostile environment where it would be unsafe for humans to go into (for example, Fukushima). Moreover, quite often, four-legged robots (doglike) are preferred because of their improved stability and agility.

The bot should be able to move alone amidst the rubble, drill a hole with a machine, close a big valve, climb a staircase, and get in and out of a car.[14] Clearly it is left to fend for itself, get up after a fall, and be able to find its own energy.

## A Robot to Replace Man

Thinking, of course, is the essential role of the brain, but a larger activity relies on the management of the body and the corresponding sensors in order to keep the human facing the permanent constraints of the environment in the unforeseeable situations life brings, alive. The challenge between the brain and the computer is not restricted to thinking, but more essentially to action. This research is mainly performed in the U.S.

---

[13] Jet Propulsion Laboratory.

[14] This especially difficult act was achieved for the first time by Hubo, a Korean robot ... but the operation lasted four minutes!

Humanoid or "animal-like" robots[15,16] were especially designed to work in rescue operations in disasters or hazardous areas such as the Fukushima nuclear plant. They have to be optimized for mobility, dexterity, strength, and endurance. They, *a priori*, are not intended to physically resemble a man (or a woman!); only the function is important.

This is not easy to achieve in a fully autonomous procedure, and "supervised autonomy" should be preferred most of the time because of the imponderables. Of course, this must also be performed in a wireless process, which raises the issue of energy and the corresponding limited self-sufficiency.

Walking and bipedal robots must benefit from extensive software that tells them how to stand up, walk, and get knowledge of the environment to foresee the action. Then they must be provided with a minimum of awareness of where and how their body is positioned. However, this human simulation quickly meets its limitations. For instance, it is much easier to make a driverless car (which is quite a done deal) than to make an android driver, which would have to learn how to get in the car, reach the wheel and the pedals ... and also get out, which happens to be the most complex operation! The car has been made for a man, not for a cyborg or a full machine, and the man does the job so simply and quickly! Even so, such a "machine" has recently passed the test. The only problem was that it took a tremendously long while to perform the whole operation — a man would have done it in a jiffy!

In the special case of a man's walking ability, it is to be recalled that the soles of his feet house a complex network of pressure-sensitive sensors which instantly inform the brain of the exact position of the body in the field of gravity and start an instant feedback in the muscles of the legs in order to compensate for any deviation from the anticipated movement. This is rather tricky to simulate in a computer; that is the reason why the bots are flat-footed and then experience so many difficulties with moving in rough terrains.

These restrictions imply that a non-humanoid solution should be preferred in any case, as long as it is not mandatory. This was the case for

---

[15] "Iron man," Lev Grossman, *Time*, June 8, 2015.
[16] The doglike robot Spot was developed by Boston Dynamics and is able to perform many delicate missions in the military as well as the civilian domain.

Cheetah 2,[17] the robot developed by MIT which is able to walk, jump over an obstacle, and recover from a tripping hazard. Four legs are better suited for walking safely!

The current robots are made to replace man in any situation where the task to be fulfilled is carefully defined and limited, and they often do the job much better than a man could, although fully unable to improvise in an unexpected situation. But, the time has not come when a flat-footed android bot will win a rodeo competition or a tennis championship.

### The Humanlike Androids

More striking is the case of an android whose job is to physically resemble a man to the point where confusion might occur.

These very special sophisticated applications are mainly carried out in Japan, with different aims. Many androids have already been proposed to take the place of a hostess at the reception desk of a museum, such as Kodomoroid and Otonaroid, which welcome you at the Mitsukoshi exhibition in Tokyo. They are able to answer questions in various languages. The resemblance is really uncanny! This innovation is quite impressive but still far from being confused with a human even though the evolution is fantastically fast.

To get closer to reality, the android must be trained to analyze the facial expression in a video signal and in real time; this new kind of software is under study in France[18] to get information on the interlocutor (man, woman, age, humor, emotions, sight direction, speech recognition on the lips, etc.).

The second step for the android, following the feelings related to the environment, would be to control the aspect of its face in correspondence with the situation. This is unquestionably a real challenge in coordinating the shape of the artificial face. Not so easy to resemble a human!

---

[17] http://portail.free.fr/tech-web/sciences/5764595_20150529_video-cheetah-2-le-robotguepard-qui-saute-des-obstacles-en-pleine-course.html

[18] Among the many well-reputed companies (Google, Facebook, and others), a startup is to be mentioned: SmartMeUp, located in Grenoble.

Currently, the robotics is not yet able to create artificial vocal cords; the voice of the android comes out of an ordinary loudspeaker. To better resemble humans, Japanese developers are trying to synchronously distort the lips of the manikin following the phonetics of the language — but the rest of the face remains desperately static, even though the eyes keep rolling in a vague direction! This is quite disappointing if the replica aims at being convincing.

A first realization in this domain is the Korean "EveR-1," which is equipped with a "facial musculature" that allows it to translate realistic facial expressions such as lip synchronization when singing. However, very recent and impressive improvements have been obtained.[19]

Other examples of "semi"-androids are more oriented toward the relation with humans, but without trying to physically resemble them. Their applications are in the field of home assistance. Pepper is a Franco-Japanese product: 1.21 m tall; equipped with four microphones, two cameras, and five touch sensors; sonars, lasers, and a touchscreen complete the panoply. A thousand bots are sold every month! Pepper is an "emotional robot," able to be sensitive and adaptive to the context in relation to an expert AI system in the cloud. It is intended to optimize the security of the nearby persons — but the company warns that it is not equipped ... for sexual relations! The bots can be met at the Softbank agencies in Japan, where they are in charge of welcoming and diverting customers (to some extent!).

Pepper's little brother, Nao, is only 58 cm tall and is dedicated to being a very interactive home companion. It is also used in software classes as a testbed for future developers.

To be mentioned also: Aibo was developed by Sony (in 1999) as a home-pet four-legged robot which recognizes its environment and obeys vocal commands while being a little bit "temperamental." It is able to learn from experience.

Then, considering all these achievements, the following question arises: if the brain, one day, were outperformed by a machine, would this be humanlike or not? Just how far would AI-driven bots outperform

---

[19]"Could you fall in love with this robot?", Harriet Taylor, *CNBC*, March 16, 2016. Available at: http://www.cnbc.com/2016/03/16/could-you-fall-in-love-with-this-robot.html

humans? As of now, all these applications work quite well individually, but the issue will be to get together and harmonize all that stuff in order to get closer to a human brain ... hand-waving discussions!

Another way to consider androids is to bypass the brain autonomy of the machine and make it directly a slave of a human brain by means of a "thought-commanded" control. A new track to explore! This will assuredly become feasible as soon as brain wave analysis is improved. Maybe this will be simpler to do than for an autonomous android.

### All of That has Been Partly Engaged But the Essential Remains Undone

Among the main attempts to get the knowledge and bring it to work, some projects are to be mentioned:

- Initially there was the project Blue Brain Simulation, run by the Swiss Federal Institute of Technology in 2005, under the direction of Henry Markram (a former professor at the University of Cape Town), but it was later changed to the Human Brain Project (HBP) at Ecole Polytechnique de Lausanne in 2013, and supported at a level of US$ 1.3 billion a year. It is a project as grandiose as the moon shot of John F. Kennedy; it involves 150 research groups, 70 institutions, and 22 countries and will continue until 2020 under the support of Europe!

  Is that the best way to address the subject? This is a tentative, blind, and bottom-up process, with mountains of data coming from cell biology to the mechanisms behind conscious decision-making; but a global overview of the behavior is also considered; the whole is put into a big computer to see what will come out of it. This controversial project of "a brain in a box" resulted in the opposition of 800 scientists, including Nobel Prize winners.

  However, Mitchell Waldrop qualified it as "a supercomputer simulation that integrates everything known about the human brain";[20] the evaluated computer power would roughly amount to an exaflop ($10^{18}$

---

[20] "Computer modelling: Brain in a box", Mitchell Waldrop, *Nature*, 482(7386), 456–458 (2012).

operations per second). In his paper no comment is made on the diversity, the adaptability, or the evolution of the neurons, which hardly behave in a reproductive way.

Makram remains quite confident and has even said: "It will be lots of Einsteins coming together to build a brain!"

- IBM is also in the running: TrueNorth is a chip (the size of a postage stamp! 5.4 billion transistors, equivalent to 256 million synapses) whose cognitive architecture is carefully modeled to mimic a part of the human brain. It is much more effective than previous circuits, with much lower power consumption. The chips can be tethered in a $4 \times 4$ configuration or even scaled up. Even though such a system is far more limited than a hundred-billion-neuron natural brain, it would still be essential to enable a huge advance in AI, producing lots of immediate applications. It is a noticeable contribution to the SyNAPSE project of DARPA.

- Google Brain: The presently most famous Brain Builder, Andrew Ng,[21] is a Chinese scientist born in Singapore. He made a major break-through in AI with a neuronal network which was able to analyze the videos on YouTube and learn to identify faces on its own. The support for Google Brain was 16,000 computers rented by Google for the experience. Now Ng is busy with an extension of this application, with Baidu (the Chinese Internet), in order to create "the world's most sophisticated neural network," with 100 billion synaptic connections, near the Institute of Deep Learning, close to Beijing. This makes us suppose that an artificial brain could be elaborated without the prerequisite of deeply investigating the human brain.

- "Connectome" is a word to specify a project which aims at establishing the whole connection scheme of the neurons in the human brain. There are many possible approaches, the craziest being whole brain emulation (WBE).

This comes from Harvard, with large support from the National Institutes of Health. The method of investigation uses a cryo-ultra-microtome to cut some micron-thick slices of a frozen brain (from a

---

[21] "Brain builder", Dan Kedmey, *Time*, October 9, 2014.

previously dead person, I guess!) in which the sections of neurons and the corresponding axons are identified through microscopic means (high resolution electronic microscopy or any scanning probe microscopic!); the information is then compiled in a supercomputer to obtain a 3D view of the whole assembly.

The idea looks rather clear at first but achieving it is not so simple.

First of all, the slicing will not reach the scale of a neuron let alone the scale of the axon section; then the tracking of all the links in the mess of images is not straightforward; from one slice to the other, the correlation is far from obvious and plenty of errors are to be expected. All of that means that, even if a complete result could be obtained, nobody would be able to guarantee its representativeness, or the use which could be made of that.

A human adult gathers some 100 billion neurons, each equipped with a lot of axons and forests of synapses; detecting with some confidence all this material only occurs in utopia. On top of that, each brain is different from another and there is nothing that says whether a dead brain is similar to a living one even from a connection point of view; even the wiring could be modified by the death process. This experiment looks more like an exercise in style than obtaining reliable scientific information on the core principles of how that system works.

How does information jump from a neuron to another (and which one?), what is its origin, and where does the decision come from? All these questions will remain unanswered, as the cerebral organization is distributed in the whole volume of the brain without a particular part being identified as responsible for a special function.

Understanding all this business seems quite unworkable. Joshua Sanes (Harvard University) said, tongue in cheek: "As a purely scientific challenge, I believe, and many people believe, this is the biggest challenge of the century."

Also, and maybe this is the most important point, the brain could not work alone; relations with close-lying organs (the hypothalamus, hippocampus, and cerebellum) are mandatory, as well as very distant organs (the spinal cord, retina, auditory nerve and, more generally, the complete nervous system). This is a dissociable set of the body's command structure.

- As far as Nick Bostrom is concerned, a copy of the 100 billion neurons could be made molecule after molecule, including their connections after cutting a brain into slices and analyzing them! This looks quite unthinkable even if already attempted. Following Bostrom, it is only a matter of technology, "even with a fairly limited understanding of the brain"!

- The same idea has been developed by "neurobotics": build a brain from the ground up, one neuron equivalent at a time (Jeff Kirchmar, 1978). Neurobots are built as a precise but limited replica of a biological system and work the same way when embedded in a specific environment; they each behave in their own way, learning through trial and error, contrary to Aibo, the Japanese robot. In doing so, the corresponding brain would acquire a specific personality. Why not drop to the level of the Schrödinger equation, the existing physics being enough without taking care of any biological model! Quite crazy but so exciting!

- To begin with a digital copy of the brain, some ways are to be considered:

  — A high fidelity emulation that has the full set of knowledge, skills, capacities, and values of the emulated brain.

  — A distorted emulation whose dispositions are significantly non-human in some ways but which is mostly able to perform the same intellectual labor as the emulated brain.

  — A generic emulation that is somewhat like an infant lacking the skills or memories that had been acquired by the emulated adult brain but with the capacity to learn most of what a normal human can.

This could be a kind of "neuromorphic AI" made up of neurocomputational functions to be hybridized with the software mockup. (I guess you surely follow this!).

No animal brain has been copied to date in spite of the many attempts already made, except that of the worm *Caernorhabditis elegans* (*C. elegans*), which only has 302 neurons. This result is very provisional even if the digital mockup conveniently animates its virtual movements. However, to improve this realization, it would be meaningful to complement it with a connected nervous system allowing a back-and-forth relationship.

The causality, the global finality of the functioning of a normal brain, looks fundamentally unreachable, even for a small animal, in a medium-term foreseeable state of our knowledge and technology. Identifying the origin of a brain animation (I would dare to say "the soul") stems from religion and theology more than from realistic science. It surely is more practical to consider a fully artificial intelligence (neurobotics) elaborated from the functions to be fulfilled without taking care of copying a brain.

### Biological Cognition

Another way to approach the problem is to imagine some biological ways to "boost" the brain! This could also be considered crazy at first glance!

A first solution to be proposed could be to play the game as natural selection does and let it go. However, we do know such a process is dreadfully lengthy and we are not to wait millennia to expect a result!

A second alternative could be to train kids: optimized food, chemistry control, sleep, and exercise, but all of this "eugenistics" seems close to tickling the demon.

The third way, which could possibly be accessible without too much trouble, could be (and this is timidly experimented with in spite of the eugenist background) to introduce genetically controlled selection of embryos or directly enter into a gene-editing process.

## What to Do, and What For?

We are in the midst of getting accustomed to machines; this is inevitable. Our daily environment becomes filled with "intelligent" or connected objects, and our relation with the machines has already changed. In the old times, one directly "actuated" the machines, and then one "commanded" them by setting the set point and watching directly the adjustments. Now, one only "supervises" the smooth running of the machine, for it has its own autonomous life; by the way, this supervision becomes more and more distant as the heater, the washing machine, and so on start working by themselves night and day, and the next day the laundry is washed, well wrung, and dried (any more and it would also be ironed!). Complete trust does reign; same thing for the pilot of the plane who "controls" his machine, sipping his cup of coffee.

## Would the Computer be Able to Reciprocally Control the Brain?

But, if such a confidence is to prevail, would the computer be able to reciprocally control the brain? Already the transfer of information brings the computer right into the brain.

I remember my early years of being a researcher just starting out. Information was then rather difficult to acquire. It was found in the library, embedded in reviews we had to explore; the presented results were often more than a year old — the delay required to write the paper, send it to the referees for review and waiting for their answer, discussing the modifications, printing the review when available and posting it to the library of the university. That was terribly lengthy. The photocopier still did not exist; one had to ask for a photographic copy (often in a remote place), and only then did one get a document transportable and achievable.

Today, things have changed tremendously; information flows at the speed of light (with the errors, the bugs, and the corresponding hasty conclusions). Communication between researchers is direct and instantaneous. The profusion of information directly and speedily reaches our desk; nothing is comparable with the old times. The computer has in some way entered the mind of the researcher; this is quite a "neuronal extension" integrated into the working process — impossible to escape! The computer accompanies the researcher at every stage; it memorizes, organizes, shows, and collects the results to send them to the community. This is quite an established collaboration.

## All of That Has Been Partly Achieved

The computer has already entered the brain through very soft processes. Cerebral training already participates in brain improvements through dedicated "games." Special programs are used to get better in mental math calculation, deduction processes, mental focus, visual attention, etc.

Professional circles such as speech therapists or psychiatrists are familiar with such assisted methods, while human resources officers traditionally use computers to help analyze personality, analytical abilities, or leadership qualities, in job interviews.

Other, more personal computer means contribute to large scale data gathering for identification through biometry: digital fingerprint analysis, face or pupillary identification, and so on. All of that allows fast, massive and precise control of the persons, for instance in passport identification. Big Brother and his young sibling Google are deeply involved in developing such tools which already participate actively in our familiar environment.

But will this intrusion of the computer into our lives lead us to a virtual reality? Are we already living in a somewhat virtual reality? That surely becomes very questionable. We would likely get accustomed to the invasive but accepted assistance of the computer in many unexpected circumstances. Is that the progress or the future we are waiting for? What would remain of the human initiative, and would we be doomed to be guided and assisted by a computer?

Or, in another perspective, would the computer be, at the end, tired of "nursing" the human and decide by itself to forge its own path and simply ignore humans? This is certainly not for tomorrow... but why not for the day after tomorrow?

# Chapter 11

# How to Make Man and Computer Cooperate?

René Descartes was the first to suggest that a *"mécanique nerveuse"* was at work in the brain, and William James[1] followed, looking for the basic principle that coordinates our thoughts.

The "evil demon" would be obsessively trying to cheat us;[2] would it be able now to creep into a computer and liven up a virtual reality which we could not separate from the normal standby state?[3]

## Is Our Brain to Change?

Our brain has become a new field for experimental cognitive psychology especially devoted to young babies' learning process,[4] which could serve as a model for the "deep learning" algorithms to be implemented in self-learning computers. New investigation methods such as MRI offer now a possibility of dramatically improving our knowledge of the involved mechanisms, but this is not without badly reawakening the specter of eugenics. In this domain we are seriously "summoning the demon"!

---

[1] *Whitehead's Pancreativism: Jamesian Applications*, Michel Weber, Ontos Verlag, 2011.

[2] *Méditations métaphysiques,* René Descartes, Paris, 1641.

[3] *Theaetetus*, by Plato.

[4] Stanislas Dehaene, Conf. "Collège de France" (Oct. 10, 2014); https://www.college-de-france.fr/site/stanislas-dehaene/symposium-2014-11-13-09h40.htm

## About the Learning Mechanism

At an early age, the organizational plan of the brain is quite similar for every baby; the following evolution depends mainly on the environment. At this stage the brain of the baby is already well structured; the cerebral areas activate as they will in adulthood. The recognition and understanding of the words heard precedes their formulation; specific zones can be visualized in a scan to ascertain if a kid knows how to read.

Some concepts pre-exist, such as consciousness of space and (especially) verticality, which is a prerecorded notion. This leads to the notion of "mirror invariance," which makes it possible to identify an identical object (or face) even if it is viewed in opposite directions. This will generate some trouble with individualizing letters like "d" and "b" (as well as "p" and "q"), which obeys the mirror effect but still refer to different meanings. To separate the meanings, educating the mental image through the graphic motor gesture is necessary. This is why it is so important to train pupils to write and draw and not type on a keyboard or a tactile screen.

The same thing is true for the notion of numbers, which does not imply any consciousness of arithmetic; calculation will have to be stimulated later by dedicated teaching. The cerebral networks of mathematics are also based on pre-existing foundations but require adapted stimulation (especially the operations using fractions which are not easily accepted). The special role of sleep in the assimilation of knowledge is also to be understood.

A baby is nothing else than an active "learning machine," even when asleep! Knowledge of the initial brain mechanisms of the baby leads to two different, straightforward applications:

• The first one is devoted to optimization of the learning process itself to achieve maximization of the biological potential. Neuropsychologists are keen to find experimental teaching methods in order to quicken and enhance the learning ability of babies and so boost their future performance (alertness, language, geometry, lectures, etc.). Yet, they are not sure of the effectiveness of their methods, or of the possible side effects of such an "imposed distortion" of the natural procedure.

However, they already envision a possible extensive use in nursery schools. Still, no guarantee is offered that the future teachers will be qualified enough to spontaneously improvise as psychotherapists of the babies. However, they assuredly will have to become qualified "neuro-players." The "neuronal recycling" will become mandatory for overwhelming the innate.

This directly leads to an improvised and extended form of eugenics, and the ideal model remains to be invented. Would it be standard or diversified into specialties? However, we still have to consider that there are limits that could not be safely enforced; there is a resilience of the brains, especially young ones. The school itself should have to get adapted to a "cognitive pedagogy," maybe using NBIC technologies.[5]

- The second application aims at transferring this knowledge to a computer to train the machine with a "self-learning" process by copying the baby's natural model. Of course, from that point on, it becomes easily conceivable to improve this initial digital version with algorithmic and programming extension in domains which could not have been reached in a natural brain. The general idea always remains to access with the machine to unthinkable specifications and performances. Another advantage of the machine is that it can be, by nature, faster but above all fully reconfigurable if required (that is hardly imaginable for a brain!). It would perfectly get adapted to the MOOC's teaching method in every kind of field.

Obviously, such an optimized "digital super-baby" should be largely more effective than a natural one and in addition it would be incredibly faster along its learning pathway, without any worries about possible induced troubles. Would it not be the best way to initiate an Intelligent Artificial Super Copy (IASC) of a human brain?

## The Brain Competition

It is well known that the competitive spirit already imparted in the educational process from the early ages until university level will be strongly boosted if the "revealed secrets" of cognitivism are placed in the public

---

[5] Laurent Alexandre, TED Conference (Paris; Oct. 16, 2014). https://www.youtube.com/watch?v=wWF7R2vs3qM

arena. A competition already held between smart people and an economy of knowledge will arise, with AI becoming mature. Bill Gates foresees that within 30 years many jobs will be replaced by automatons, while Ray Kurzweil says that around 2045 AI will be a billion times more powerful than the brain!

This especially holds if "thinking and learning" machines are to help retrofit the human brain into boosted new humans. Who would be reluctant to give his or her kids the best conditions to confront the realities of life and the corresponding ruthless professional competition? We are already in some ways a "machine to win" to make our living and that trend will surely be exacerbated in the future. Society will no longer accept the inequalities in IQ, and solutions will have to be found to fill the gaps; nobody can be left behind and the intellectual eugenics will be accepted.

Would that be accessible to everybody? That is another question. Slow learners will certainly have to sink or swim, because the economy will certainly increasingly require advanced and updated knowledge!

This race to the performance is even sustained by geneticists, who propose, at the very beginning of the chain, a genetic sorting of the embryos in order to select the intellectually most promising ones. It is already known that the genetic part of IQ (50%) is more important than previously thought, the remainder being attributed to education, family, and school. Such an active selection is already being busily experimented with in China using selected high IQ donors, and this has become an accepted eugenics similar to the previous versions, which have been so vigorously discredited in other times.

Competition already occurs in every field of human activity, but it is especially exacerbated in those fields where important and spontaneous financial opportunities may exist. These circumstances lead to blameworthy abuses such as corruption or mafia organizations.

Today, one striking example comes from sports, where the individual ambitions and major financial stakes involved have a crucial bearing; the practice of doping has undoubtedly become omnipresent. It is accepted and even strongly supported by athletes whose only dream is to conquer the podium, whatever the price to pay. But, also, the corresponding organizers and sports companies find a way to make money from the rigged competitions.

The means implemented to detect and fight these abuses find immediately a fully effective technological defense; this to such an extent that, today, the winner is likely the one who is the most "improved," whether individual (a famous example is Lance Armstrong) or collective sports are concerned.

These abuses are so pervasive that they quickly spill over into the geopolitical domain at a world scale (an example is the current scandal involving Russian athletes).

The technical innovation plays a powerful leading role in the domain of advanced industry with all the huge economic stakes involved in the competition: the pharmaceutical, electronic, and aeronautical industries among many others, with no holds barred in dominating the market.

What will happen when the "improved man" emerges on the public stage? Whatever the improvement involved (brain, muscles, health, longevity, etc.), that will assuredly trigger a strong economic resonance, because it will be at the world scale, with enormous markets that the companies will fiercely fight for.

The markets are of two types:

- One is related to the "production" (maintenance and updating) of the best "improved man" (which certainly will not be for free even if it comes from Google).
- The other is related to the "use" of these improved colleagues who are intended to ensure enhanced productivity with an exceptional yield.

At this point, it is safe to imagine that these high stakes will draw in interests, actions, and also corruption, especially perverse kinds coming from the international mafia organizations (an example is the current soccer scandal).

How far could we go and what timing should we have in order for these innovations to be gently introduced so as not to unsettle our societies and to prevent an excessive impact?

## The Hybrid Thought

All of this clearly shows that we have begun to seriously "put our fingers in the brain." We are proceeding gropingly, without exactly knowing what we are to do, but that could be considered provisional as responses are made, little by little, to the questions. We are already aiming at deciphering the very origins of thought and trying to deduce models which can be digitally transcribed. With such an improved knowledge it is expected that dedicated digital implants could be conveniently installed to enhance the brain's activity, compensate for deviations, keep a relation with remote computers, and so on. Then a "hybrid thought" could possibly be developed and shared directly with computers.

It is to be recalled — that neurons are some million years old and the transistor only 60, so let time do its work! Ray Kurzweil, the Pythia of our times, imagines that within 20 years brain implants will be in common use to connect us with the Internet. Hybrid thought will come with Humanity 2.0.

## The Biological Copy

Basic questions arise from this idea of achieving a copy of the human brain: Why a brain-like copy? To what extent "like"? A live copy? Would it not be better to try to directly improve the brain? What to do with a copy? Also, second order questions come: What does it mean "to think"?

What about limiting unpleasant sensations — pain, stress, fatigue? What about the pathological disorders to be escaped? What is a sane, a normal, an exceptional, a crazy, a dangerous brain? What kind of brain is to be preferred for a copy? All of that comes from the new theory of "computationalism" supported by Hans Moravec and Ray Kurzweil;[6] a quite strange word used to define the idea that "the human mind or brain (or both) is an information processing system and that thinking is a form of computing".

### *Making a Direct Copy of the Biological*

Could this be done for real? We have already invoked earlier this endeavor, but some discussions have to be completed and we have to deal with it again.

---

[6]E.g. "Life" Edge.org, John Brockman, Ed., 2016.

If such a project could be possible in the future an immediate question arises: who is to be chosen to be copied? Would he be a prominent scientist, a philosopher, a famous artist, or any kind of volunteer (particularly if his brain has to be sliced like a ham before being copied![7]). To be an exact replica, this would require also copying the same memories, the same experiences and know-how, and the same story.

It has also been recently shown, using MRI scanning, that cardiac arrest does not stop the brain from remaining active, albeit in a weak and disordered manner for a period of up to 32 hours. Nobody yet knows the wiring changes which this process could have made in the brain before it was "sliced."

Whole brain emulation (WBE) is quite fictional a project. Even a low level brain model (for instance an ant or a worm) involving a very limited number of neurons exhibiting elementary rustic functioning will require tremendously important efforts.[8] Of course, this is not concerned with a dualistic and mystic soul and spirit philosophy.

The starting point of this idea is to establish a connecting map of the neurons even at the molecular or atomic scale (atom after atom and possibly invoking the Schrödinger equation![9]). For the sake of completeness, the same operation should be extended to the spinal cord, the sensory receptors, and the muscle cells.

Such an endeavor would certainly imply knowing in detail the exact role of the DNA, the proteins, the neurotransmitters, and the hormones; will we have to simulate all of these molecules or even atoms individually? This actually is a "top-down" framework of reverse engineering!

All of that stuff looks absolutely mad to a sound-minded physicist! Then we would have to reconstruct in a computer the three-dimensional wiring of the neurons. To proceed we obviously would also have to know exactly how each neuron works (they all are genetically independent) in order to make a complete virtual mockup of the brain. Eventually we would have to explain to this artificial brain what game it is to play.

---

[7] This is not a fantasy; it is really emphasized, as we have seen.

[8] Such a low scale project has already been launched at the Janelia Farm in Virginia, under the leadership of the Howard Hughes Foundation. It is dedicated to copying the brain of a drosophila (fruit fly).

[9] Böstrom, *op. cit.*

This is even though, as we have seen, such a tentative reconstruction has already been undertaken on the worm *C elegans*. To know if the computer simulation could be relevant, it just remains for us to determine how the worm thinks and behaves! This, surely, is not for tomorrow, even if only 302 neurons (Ch. Koch) are involved. We are putting the cart before the horse.

Ten years ago, Henry Makram[10] was initially very pessimistic about this research, but the success they have obtained in the simulation of a part of a rat brain now makes him more confident of finding a solution within the next ten years.

Replicating the whole neuron assembly with microchips would require a tremendously large amount of material. If the cortex-like columns of chips were assembled in a hyper-column structure allowing sequential operations (as the neurons are), then the speed of the electronics would be slowed down to such an extent that it would be surpassed by the biological challenger.

Despite all of that, many people trust in the feasibility of such a crazy mechanistic project[11] on IBM's supercomputer Blue Gene, even at a humanlike scale! Not any mention is made of reconstructing the life process or the corresponding consciousness.

### The Complexity

Let us again remember that in a human brain there are 100 billion neurons, each equipped with at least some 10,000 synapses, not to mention the countless "glial cells," the precise role of which is still to be discovered. The model also has to include the knowledge of the chemical mechanisms of the ion transfer which make chemical substances (hormones) cross the blood–brain membranes, and many other protein interactions. That's a lot of subtle things that would have to take place in a real

---

[10] Director of the Blue Brain Project. See "Artificial brain '10 years away'", Jonathan Filders, *BBC*, July 22, 2009. Available at: http://news.bbc.co.uk./2/hi/8164060.stm

[11] Anders Sandberg and Nick Boström (2008): "The basic idea is to take a particular brain, scan its structure in detail, and construct a software model of it that is so faithful to the original that, when run on appropriate hardware, it will behave the same way as the original brain." Of course, that's obvious, Dr. Watson!

time simulation and taking into account a random noise environment contributing to perturbing the signal transmission.

On top of that, the question remains on the way the "virtual embodiment" could be reached to complement a basic meaning. Not to mention the requirement to be connected to a representative "3D body" (decentralized system) providing the right information about the surrounding world and also sometimes contributing to the subcontracted treatment of specific information (such as the spinal cord). Let us say that this world could as well belong to a virtual reality or a simulated world.

Such a "mind uploading," if it was to become possible, could be of significant help in extreme situations; it could be considered as providing a backup of the "mind file" that could help to support who we are in case of a disaster or to withstand lengthy space travel. But the question remains: who would be there to press the "ON" button on arrival? Could this file allow us to initiate a 3D printing of our brain? We have just entered pure science fiction.

Of course, such a project would require a very long delay to carry out (if it could be). Many other general or partial conceptually different solutions could have arisen beforehand.

In spite of all these huge problems and difficulties, the WBE project helps to facilitate partial solutions of subprojects which are of considerable interest in many other applications, such as the analysis of large neuroscience datasets or corresponding theories. The way the research is focused also helps researchers coming from different fields communicate.

A similar effort is undertaken in the U.S. under the title "BRAIN Initiative," but it adheres to a different structure of organization. In both cases it is not a matter of science but of advanced technology. But the more technology helps to discover anything about the way the brain operates, the more we can understand its underlying nature, and this is no doubt science.

## Software Brain Emulation

Kurzweil promises that a brain transfer will be achieved in the near future. Of course, before that we will know what is to be copied; that is, we have to understand exactly the basic principles of brain's functioning, which are far from being acquired. At the moment, the brain is still an unfathomable mystery even though the progress of knowledge is fast.

## Low Level Emulation

However, a high level cognitive AI copy of the whole brain is not required to achieve partial brain emulation; the low level structure of the underlying neural network could be identically emulated in a computer system if the operational principle and the interaction of the neurons were conveniently deciphered. The parallel computation is a natural fit for the neuronal processes.

This approach, in some way disassociated from the biological brain, is by far more realistic and can rely on partial software solutions as they become available. Only the functional aspect is considered and, as a matter of fact, this has been implemented for years in lots of practical situations, sometimes quite complex, where the computer has proven to be largely more suited than humans regarding speed and reliability: expert systems. For instance, weather previsionists do not care about the exact position and evolution of each drop of water in the atmosphere; only the global behavior is required (of course, weather forecasting is not an exact science).

In any case, if a brain copy is to be achieved, whatever the substrate, would it be related to the original brain in real time in such way that any change in the thoughts would be immediately replicated in the copy to keep it updated, or would the copy be free to evolve on its own? In this last case, strong divergences would occur very rapidly and the copy would no longer be a copy but a self-minded "entity," especially since it might not inherit the same human motivations.

## Are We Living in a Dream?

Hilary Putman is a "mathematician–philosopher"[12] in the purest tradition of the Greek thought that develops research on the knowledge of knowing oneself. Plato already in his famous *Theaetetus* asked the question of how

---

[12] Putnam, *Reason, Truth and History*; *The Threefold Cord: Mind, Body and World*, Hillary W. Putnam, Columbia University Press, 1999.

to distinguish reality from a dream. Same thing with Descartes and his vile demon leading our illusions about the real world, and now Nick Bostrom claims that our universe would be nothing else than a computer-like simulation designed by an unknown intelligence. Obviously, we have no specific means to assess the appropriateness of these daring and purely academic hypotheses.

Putnam is a philosopher of the mind who developed several successive theories which he soon contradicted, so demonstrating that nothing is to be taken for granted in this domain even if one accepts the reality of the mathematical entities.[13] The conception of a human brain, the philosopher says, is expressed as an equal competition between "functionalism" and "computationalism." Humph!

In another way, Putnam is also famous for asking the question whether the brain is right to believe in what it trusts in. He elaborated on the paradox of the "brain in a vat," imagining an active brain placed in a vat and provided with all necessary means to survive, and artificially stimulated by a computer for external virtual information similar to that of a real body. This is the ultimate version of virtual reality. In such a circumstance, what could be the free will of this brain? How can we be sure we are not, ourselves, such a brain in such a vat? This is the "hyperbolic doubt" of Descartes: where could the evidences be? Many films or videos have reached for this theme.[14]

Is our vision of the world an illusion we are accustomed to since our early days? Is it purely conventional and not directly related to an exact reality? Would Life and Death be only an illusion? Who can assure us that the perceived reality is the true reality? Illusionists excel at creating such misleading illusions.

## The Virtual Reality of Our World

Michael Abrash[15] is a video game programmer involved in the creation of purely virtual environments (sight and sound). He said: "Reality is an illusion. Reality is not what does exist but only what your brain records."

---

[13] It is worth observing that such "entities" always rely on a material reference.

[14] *Matrix, Dark Star, mondes possibles,* etc.

[15] See https://en.wikipedia.org/wiki/Michael_Abrash

Current means easily allow fooling the senses with subliminal images, judders, and other such perceptual artifacts but there's the rub: sensorial hacking!

That's a reasonable take right now, but we think you should pay close attention, because VR is likely to have a big impact much sooner than we could imagine. Decoherence between image and sound can cause confusion in the words.

The key to making the experience unique and compelling is convincing perceptual systems that operate at such a low level — well below conscious awareness — that they're perceiving reality. This is a common practice in the movies which import the real world into the virtual one.

This issue of VR opens up the question of how to directly influence a brain from the outside, how to persuade somebody of an external and artificial reality, how to command a brain. Hypnotism has been going on for an awfully long time without anybody being able to explain how and why it works so impressively.

Will electromagnetic waves be capable of any reciprocal interaction with the neurons? We don't yet know. However, there are people who claim to be disturbed by the radiated power of WiFi, mobile RF transmitters, or nearby high voltage power lines. Nothing is clearly ascertained or biologically measured. Nobody knows how the brain could be involved. Be that as it may, the day we learn how to transmit information directly into the brain, that should be a breakthrough.

### Could the Computer Invent a New Model of Brain?

The following question immediately arises: What for? There could be two different objectives:

- The first one is to achieve professional improvement. Our current way of life drives us toward constant updating of our capacities in order to become more competitive and effective. If a machine was able to help in this quest for promotion, we could not remain indifferent.

  A particular application of implementing new kinds of brains is related to space travel, especially long ones (to Mars).

- The other possibility is simply a search for its own satisfaction. Not to mention that the selfish and hedonistic will be different and special, such as sporting a tattoo or a piercing!

Of course, even if an improved version of the brain is to emerge from a computer, the question will remain of how to transfer it back into a skull. It would certainly be better not to care anymore about human improvement and to let this intelligence be in the silicon world, where we should know how to put it into action (or stop it)!

## Androids and Cyborgs

Would machines be worse or better than humans? Man has largely demonstrated for all eternity that he is mostly unable to live in peace and freedom. In this respect, would it not be preferable that machines not be built on a humanlike model? However, the machine is, for the moment, one up on the human: it only obeys preset software which can be changed as required (or bluntly cut off, such as HAL[16]). There is no need to make war and kill any computer!

However, the fear remains that if the computer knows everything, if it becomes intelligent enough to understand the human world, if it achieves its own personality (and this could be on the verge of happening), then why would it not take a big bite out of life and gain power, as a human would certainly do? He[17] could possibly get the opportunity to do it.

### Androids

First of all, what could be the social role of such humanlike mechanics, and what would be the requirement for being humanlike? In most of the practical applications, the required field of action is restricted to a precise function which does not need the elaborate and diversified potentialities of being fully humanlike: expert systems exist and their performance is

---

[16] The famous ultimate computer from the movie *2001: A Space Odyssey.*
[17] Sorry, I typed "He" unmindfully!

fantastic within their domain, and they do not need to become full androids.

However, there are also applications which require a very "human talent" and appearance, such domestic helpers to seniors, reception staff,[18] even workers or cooks. As long as such polyvalent androids can be achieved, nothing would prevent series production (like for cars) of an army of dedicated servants.

The android has strong advantages over humans in many situations (it will have to charge occasionally similar to human sleep). The android does not have an own will or destiny, it just does the job, it can work day and night, it is assuredly replaceable and duplicable, it will never belong to a trade union, it has not any special need except for electrical power.

An alternative for androids is to make them slaves of humans as another body in order to avoid long or unusual work, to be present in several places at the same time, and to perform dangerous work (such as in war or policing). This could be especially profitable when a direct command by thought becomes available. That is beginning to be conceivable, and sophisticated applications may follow. It has not yet been thought about seriously, but it won't be long before it comes.

Now, what about the present state of the art in android development? Several significant outcomes already exist, and the Japanese are the most advanced in this domain:

- Asimo,[19] from Honda, was a pioneer in the year 2000. It is 120 cm tall and is able to dance, or run at a speed of 3 km/h! It is effective in vocal and face recognition but its main specialty lies in its exceptional manipulative skills.
- As mentioned before, Pepper,[20] a humanoid robot, was created by a French laboratory with a Japanese contribution (Aldebaran, 2014). It has the height of a five-year-old child; it is covered with sensors, lasers, sonars, and other amazing gimmicks, which enable it to find its

---

[18] This is the reason why most of the realizations of androids simulate girls.

[19] Advanced Step in Innovative Mobility.

[20] This name comes from a French joke: a man who is regarded as quiet, friendly, and good-humored is usually called a "pépère." See: 01net, No. 817(2015).

way or memorize its working environment. Also, its elaborate AI allows it to identify its contact person with its face and speech recognition software and understand if its counterpart is smiling, angry, or well disposed; it is allowed to detect an emotion, for instance. From then on, its answers will be adapted for a free conversation. Of course, its intelligence is cumulative all along its experience and could be transferred to a chip of it or to a memory space in the cloud.

Pepper is an extrapolation of the first robot — Nao (2006) — by the same company, which was widely distributed for years in a lot of applications. Nao is only 58 cm tall and is provided with face recognition, vocal synthesis, tactile zones, etc. Mainly used in research laboratories and as an educational platform, it is famous for its interactive skills with autistics or Alzheimer's sufferers.

- As has been mentioned before, on June 27, 2014, the National Museum of Sciences in Tokyo introduced two new cute hostesses named Kodomoroid (teen) and Otonaroid (adult woman) who are in charge of welcoming and informing the public. These graceful creatures are no more than androids surprisingly resembling human girls. During a press conference, Kodomoroid played a journalist reading the latest news; whereupon it appeared that, carried away by this success, it would like now to present a TV show of its own!

Then, in the domain of androids, the sky is the limit!

## A Mix Man/Machine: The Cyborg

Indeed, as we have seen, our knowledge of the brain and simultaneously the improvement of its simulation in the computer are both progressing very fast, each in its own way. What could already be the overlapping ranges?

The scientific discipline neuropsychology seeks to provide an answer to the *mécanique nerveuse* of Descartes,[21] and its simulation in neural networks is under continuing improvement. Artificial neurons are also improving, but there is still a lack of knowledge about how to teach them

---

[21] *The Principles of Psychology*, William James, H. Holt, 1890.

to live (even artificially). The flexibility of language has so far been a very complex obstacle, but this is on the verge of being overcome.

The internal organization of the subconscious remains a mystery but is of a key importance in the functioning of the brain. There is very little knowledge about the involved hidden mechanisms; however, some procedures for data restoration or their mass processing or their probabilistic weighing strangely resemble what a computer does.

The day we know how things are arranged in the depth of our minds, a relevant simulation will be thinkable. At the moment, it remains easier to let the subconscious work freely and only copy what is reachable. The "flexibility" of the brain comes from the subconscious and is especially required in the setup of the language, for instance, and in such domains machines are progressing quite fast. However, the way things are analyzed in a conscious mechanism is still unknown and inaccessible to a computer. Then how do we conceive a mixing of the best specifications of the brain and those of the machine in order to achieve better performance of the "assembly"?

There are two ways (which somewhat overlap) to consider this issue: the first is to improve the brain by implanting circuits which could directly collaborate on the brain's activity, possibly with external assistance; the second is to improve the machine through a brain contribution, possibly with internal assistance.

Of course, this collaboration in a first step would be restricted to specific and limited domains of thinking or action, waiting for a future extension to larger areas, the difficulty being to know the exact locations of the active zones and the way to stimulate or detect them. It remains not so easy to electrically "speak" to the brain!

Then a direct intervention in the brain through an implant to bring it what it lacks has many practical difficulties, as we have seen. Other, non-invasive techniques should be preferred.

This kind of operations has already begun for a while, especially with the improvements in the techniques of thought command through the detection and analysis of brain waves. The direct way back for information to transfer from the machine to the brain remains in its infancy and only the classical vocal or visual exchanges can be managed, as nature allows it. Then, also, the computer has to be made somewhat conscious to reorganize the information to be transferred. The first attempt to make the computer aware of something has just been successful with the robot Nao, as was mentioned in Chapter 7. This is just a beginning, showing that this will be feasible to some extent. If the computer could be conscious, then it would be able to think and the poet would be given an answer to his question:[22] "Inanimate objects, would you have a soul?" The day the computer has a soul or simply understands, "he" will be![23]

The progress of computer simulation of the brain linearly follows the progress of the computers in power, memory size, and speed of calculation. The possible scenarios for this development remain quite fuzzy but will become clearer little by little.

Would it become possible to directly operate on the brain to bring it what it lacks (to be defined)? This operation could be biological or mineral through implants — or why not through downloading?

It can easily be foreseen that all these innovations which are still to appear will not be made all at once. Many tests and trials, some unavoidable failures or mistakes, changes, prototypes, and beta versions will occur before a provisional solution becomes accepted, and before an extension to a large number of individuals becomes possible. There is not any sudden "singularity" to be expected in such a domain, but likely that will take place progressively, according to the technological advancement.

In a first step, most humans would not be as concerned about these "improvements," as was the case of vaccination, for instance. They would likely be reserved for a selected elite (at their own risks), as has already happened in space exploration.

A key point will surely be the communication with the Web and its "data mines." Currently Google does not understand anything other than keywords. Things will change when it becomes able to understand questions and enter into a discussion, and that will come soon. There is a hierarchy of difficulties to overcome and the basic question remains: what exactly is meant by "to understand"?

The Turing test starts to break down but the intelligence field involved in a robot remains limited.

---

[22] "Le lac", Alphonse de Lamartine, 1820.
[23] "I think, therefore I am," said Descartes.

## *Is Big Brother Already Here?*

Many smart people have been concerned about the risks of the fast advances in science and technology in the last few years. These new skills promise to bring remedies for all our ills and also to extend our lives beyond the current limits, but, at the same time, some worrying aspects cannot be discounted — in particular, drastic and inescapable changes that will occur in our way of living.

A threat would be the creation of a superior intelligence that could, in the near future, rule the whole of humanity in the manner of Big Brother and limit our freedoms through a superior, rational will. Would that not be, to some extent, sometimes desirable?

Actually, we do live in such a position of dependence we have created. We got accustomed to it and we did not realize that the changes were there in our daily lives and that the conditioning had taken place.

Computers, on one hand, have got a superhuman dimension and biology; on the other hand, they allow unthinkable innovations which are upsetting our conceptions. All of that has resulted from the current means of collective indoctrination through TV and the Internet. Deeply entrenched lobbies can create a people's morality which cannot be individually opposed. The current doctrine leads to deliberately disputing the ancestral natural laws in order to just do whatever we want. We no longer accept that nature dictates its laws; we want now to impose ours.

For a long time, we have been accustomed to "transgressing," for instance in taking care of ourselves, in vaccinating, in saving the lives of premature babies. But that is not enough. Now we want to remake the world in our way, from the bottom up, and biology actually allows for unexpected innovations.

Nature (some say God) has made diverse humans; one does not know why. A Chinese does not resemble a Papuan, a European differs from a Melanesian, a Bantu from an Ethiopian, and so on; same thing in the whole of the animal kingdom, which we belong to. This diversity was once called "race." In the U.S. (which is a real melting pot of races), it is well known that blacks are gifted in playing jazz whereas whites are gifted in country music. It is up to you to make a choice, but this genetical discrepancy is in no way discrimination; each person has his own talent.

Today, the public's morality protests at this designation, which nonetheless reflects reality. They say there are no more races and that biology will help us to remove these differences. We will create life on a standard model and everybody will lose his own character. What to do about races if man is to be copied in a machine?

Likewise, God (some say nature) made men and women different for procreation. Again, unacceptable discrimination! Now we are (almost) able to create life *in vitro,* in a cellular way; this difference will no longer hold. We do not accept nature; we are going to reconsider it our way.

But beware the unpredictable or even violent reactions of Nature through not yet properly controlled creatures it keeps in store; I mean, for instance, viruses!

Then we do not need to wait for Big Brother; we are already able to play God. All of that has already taken place and will only increase whatever our views; Google is here to help.

### A Bit of Science Fiction

Ray Kurzweil, our champion of transhumanist (or posthumanist?) previsions, recently forecasted, in a visionary delirium "In the 2030s, we are going to send nano-robots into the brain (via capillaries) that will provide full immersion virtual reality from within the nervous system and will connect our neocortex to the cloud. Just like how we can wirelessly expand the power of our smartphones 10,000-fold in the cloud today, we'll be able to expand our neocortex in the cloud."

"Let's digest that for a moment," commented Peter Diamantis.[24] That would (please note the "would") mean that:

- Brain-to-brain communication would become an ordinary reality. That would also mean that privacy is over. That's it. But you might know anything you'd desire.

---

[24] "Ray Kurzweil's wildest prediction: Nanobots will plug our brains into the web by the 2030s", Peter H. Diamandis, SingularityHub, October 12, 2015. Available at: https://singularityhub.com/2015/10/12/ray-kurzweils-wildest-prediction-nanobots-will-plug-our-brains-into-the-web-by-the-2030s/

- Intelligence would be scalable: your computational power would be that of a supercomputer.
- We would live in a virtual world, your brain being driven by an artificial gaming engine.
- Our improved immune system would keep us healthy thanks to a nanorobot which plays the role of the current T-cells but better.
- Any expertise could be freely accessible from the cloud, and any competence available at a click of the brain!
- Expanded memory would allow remembering everything, because everything would be stored in the cloud.

Kurzweil strongly believes that a connected neocortex through cortical implants will bring us to a higher order of existence which will change everything (for the best and the worst?).

# Chapter 12

# Is Transhumanism a Realistic Future?

The whole world buzzes with comments[1] about the topic of "transhumanism" or even "posthumanism" — that is to say, a likely definitive transformation of the human being through the emergence of new technologies in the near future.

For some people, the change could be sudden and very near, as a rupture in our way of life and they call it "singularity" (from the corresponding mathematical term which effectively designates a discontinuity in the evolution of an algebraic function). Must we have confidence in these predictions? Would it be better say "singularities" if the changes are to affect sequentially different fields of interest?

The driving force of such a change is obviously the economic investment (overt or hidden) and the corresponding expected returns. Questions follow: who will pay, at what amount, for what purpose? Nothing is free and everything is governed by an intention. Would we be able, collectively or individually, to oppose this change or would we have to accept it blindly? Would that be just a "canard" launched by some lobbies with an obscure intention?

The whole society is involved in its various and relatively independent segments: military, industry, business, and others. This also entails national or political involvement in such a no-return situation. Not easy to follow the previsionists and believe in such a harsh revolution!

---

[1] Fillard, *Transhumanism.*

# Where do We Come from, Where are We Heading to?

## The Place of God in Humanity

Gods are an intellectual creation of the human spirit eager to get an eschatological explanation of his origins and his fate, an explanation to help to soothe the rough life of the times.

From time immemorial, man was totally unaware of his surroundings. Everywhere was pain, need, hunger, danger, fear, and discomfort. As soon as the consciousness of his existence grew in his primitive brain, the existential, inescapable questions were raised: "What am I doing here, standing on my two feet? Where did I come from? Would I have another destiny than waiting for my death? Is there something after death? Where did this world come from?"

He then got the idea that he might be the product of superior wills dictating everything in this lowly world. He soon considered that he was the only using common sense or maybe even intelligence. Then he imagined, in his own image, some still more superior and everlasting being(s) possibly able to be at the origin of such an incredible world. He gave them the common name "gods." This procedure was absolutely logical; religions were born, as can be said, naturally, as a direct and effective consequence of the emergence of consciousness against a background of individual anxiety.

In a first step, these gods were put everywhere in order to satisfy all of the mysterious and disturbing observations: the seas, the mountains, the clouds and storms, the rivers or fire, for instance and, of course, the sun; all the elements he was in contact with and which, depending upon circumstances and the whims of the gods, might turn out to be either benevolent or hostile. To earn the favor of these virtual entities, this primitive man then imagined all kinds of ritual ceremonies, sacrifices, and gifts but without being able to verify if the recipe was a good one. This mythological elaboration was obviously accompanied by an adapted ceremony, a cult, and a "catechism" in order to structure and ensure the transmission and the sustainability of a common belief.

Then all these diversified conceptions were concentrated in a single almighty entity (under various terms, depending on the place) and at the origin of everything: God. Things became simpler to manage and more credible for the outsiders. Because man did not know anything other than himself he, quite naturally, gave to this supreme entity the same attributes as man, because he considered man to be the only creature worthy of getting in contact with God.

Consequently, God was pictured as an old man with a long white beard who commanded respect and embodied wisdom and timelessness. Things changed when man discovered there was nobody atop the mountain or in the sea, or in the deep skies. So God is currently expected to be a kind of alien in a remote and different universe somewhere beyond quantum mechanics and relativity. The Big Bang is emphasized, but what was before that? Who created God? The questions are endless, inconceivable!

Theologians associate the Big Bang with the creation of the world presented in the Bible; they argue something must have triggered this explosive event and only God could have done that. Such a theory, called "quantum gravity," could give a complete description of the world, from the subatomic scale to the astronomical one. That remains to be seen!

Religions still play an essential role in people's lives, on the one hand at the individual scale to provide them with "peace of mind," guidelines on conduct, a conventional ethic, an expectation of a later life less uncomfortable than the real, harsh world of the old times; on the other hand at the collective scale because religion was (and still is, provisionally) the only means of enabling people to live in freedom (and hopefully in peace), with common and imperative convictions.

This way of doing things has been quite perfectly accepted and implemented.[2] However, faith induces a force so persuasive that it is sometimes difficult to master it, and that leads us to frightening atrocities. The religious grip on the human mind can be devastating. So, would the computer fall into similar unmanageable excesses if taught religion? Such a question cannot be discarded.

However, despite such violent excesses throughout the centuries (until today), religions ensured the survival of societies. Some disappeared; some emerged, maintaining the basic principles.

---

[2] Richard Dawkins in *The God Delusion* considers religions as "memes" or "mind viruses".

Then, little by little, there came modernity, bringing a better knowledge of our surroundings, and a better mastery too. This was sometimes in contradiction or even conflict with the groundless convictions previously implanted in minds. It has been necessary to find compromises to make the ancient convictions coexist with the new discoveries.

Even though a deep mystery exists, some are convinced[3] that there are good reasons to assume that science will provide us with a full understanding of the universe that leaves no room for God. However, Sean Caroll concedes that "the idea of God has functions other than those of a scientific hypothesis." Such beliefs in the supernatural and the afterlife also play the role of societal glue and motivate people to stick to the rules. This brings us back to the very origin of the religions and the fear of death.

Today, we all know that our world (and we too) is made up of matter (atoms); one knows that there is no old man on top of the mountain and that the sky is (almost) empty. The Big Bang is still a mystery that theorists are trying to explain in vain, and dark matter is still invisible even though we know it exists somewhere in the infinite universe. No one knows what the simple notion of "time" is for, although it governs our lives. Of course, God is not made up of atoms or subject to our physical laws — then what does He consist of? Does He consist of something conceivable? Is He insensitive to time? If not, who was before God? Science is still not able to answer such basic questions, like what is "before" and what is "after" in a world where the notion of time has to be created.

How do we teach the computer about such things which we are not able to dwell on?

When the first machine appeared, in early 1944, people in the English-speaking world decided to call it a "computer," but that did not suit the French scholars, even though the word "computer" unquestionably comes from a plainly French word out of the Middle Ages: "comput".[4] So these scholars desperately searched for a drastically different word

---

[3] Sean Caroll, a cosmologist at Caltech: "As we learn more about the universe, there is less and less need to look outside it for help." This, of course, is a totally gratuitous statement which can obviously be challenged.

[4] Comput means to calculate the dates of the Christian holy days in the calendar.

consistent with French minds. The Latin philologer Jacques Perret proposed *"ordinateur"*, which literally means "God puts the world in working order"! This shows that, from the beginning, God was to play a role with computers!

## Is God to be Forgotten?

Technology promises to guide our lives into a new, "posthuman" future. Clearly, the advent of computers and our quest to make them intimately cohabit with humans (if not to replace them) lead us inexorably to raise the following question: what to do with God in this context? Is the very idea of God of any interest (individual and/or social)? Is man ready to forget the idea of religion which up to now has led his destinies, and thus return to the ignorance and fears of the early days? In such a case all the efforts spent during the millennia to elaborate the sciences would have led to nothing. What a depressing conclusion!

However, even if this could be considered surprising and somewhat contradictory, the achievement of modern science emerged from Christianity, although the very roots of rationality originated from the ancient Greeks, before Christianity.

Unquestionably, we have currently reached a fundamental crossroads. Transhumanists propose a new man to come soon, more intelligent, more adaptive, aging-proof, harmoniously cohabiting with more and more intelligent and powerful machines.

Would our universe now no longer need God, waiting for a more precise answer to the very reason for our being? Some people currently ask: are we now playing God with our exploding scientific and technological capability? But, fundamentally, do we really need to explain everything[5] in order to feel serene in facing death?

The religious constraints have currently been replaced by other constraints related to the "social consensus." No more wild beast before the

---

[5] A French poet gave us this marvelous verse about the death of a wolf:

*Fais énergiquement ta longue et lourde tâche,*

*Dans la voie où le sort a voulu t'appeler,*

*Puis, après, comme moi, soufre et meurt, sans parler.*

entrance to the cave, but the daily professional stresses have taken their place with the requirement of competition at school since an early age. The computer has already dehumanized many aspects of the daily life and pushes man into individual, solitary isolation, as we can see from the smartphone virtualities.

The computer does not have, *a priori*, any need for God, even if we try to teach it. Some psychologists are expecting to understand how the idea of God is conveyed in the human brain. One begins to better understand, through "neurotheology,"[6] how the visual and acoustical (infra sounds) environment induces in the brain meditation or emotion and how the subconscious is able to awake to the idea of God. However, no "God spot" has been discovered to date between the neurons.

A more frightening hypothesis would be to raise the question of knowing whether the computer could, in some way, replace God in the (weak) minds of some humans through an overstretched virtual reality. This hypothesis remains to be considered, but this would suggest that (super)-computers previously took a grip on our destinies.

In relation to the idea of God and always with the same aim of providing psychological relief, the concept of the "soul" has also long been developed: a virtual entity individual to each person,[7] which is defined for all eternity and destined to endure after physical death. The soul is supposed to be influenced through our behavior during our lives on earth, and religion persuades us that our soul will bear the consequences of our turpitude beyond our death. This will stimulate men to behave well; if not it is the way to hell!

As a matter of fact, we have vainly looked inside the brain for any presence of a reference (material or biological) related to the soul. The cerebral "plasticity" opposes any permanent neuronal organization; everything is mobile, modifiable, and evolutive. The "personality"[8] itself is a fluctuating notion that is somewhat elusive.

---

[6]*Neurotheology: Virtual Religion in the 21st Century,* Laurence O. McKinney, American Institute for Mindfulness, 1994.

[7]Nothing is ascertained for animals.

[8]A notion which could be considered the closest to the soul or the self.

It could be possible to associate a kind of personality with the compuater after a heavy "deep learning." Such personality would essentially originate from its "acquired" and automatic responses but with regard to the possibility of revealing a soul in a computer...that has remained pure fiction to date.

However, Kurzweil, for his part, foresees that within a couple of decades a superior intelligence "will dominate and life will take on an altered form that we cannot predict or comprehend in our limited state." At the same time, for sure he says, the brain will be downloaded in "neurochips." For the first time in evolution, technology brings us the means of controlling our destiny and "God has competition"![9]

Stephen Garner has also been writing about an audacious extrapolation: "The essence of a person can be separated from their body and represented in digital form — an immortal digital soul waiting to be freed — an ideal that some see as medieval dualism reincarnated."[10] Hans Moravec, too, thinks that intelligent machine (*Machina sapiens*) and mind transplants could give us immortality![11] Others are shouting "God is dead!" Do they mean God or the idea of God?

Are we starting to fall into a kind of intellectual delirium? Teilhard de Chardin,[12] even if he was fascinated by the new technology, already echoed the present concerns: "A scientific justification of faith in progress was now being confronted by an accumulation of scientific evidence pointing to the reverse — the species doomed to extinction." Of course, at that time, he was not aware of the Singularity of Kurzweil, but he nevertheless was speaking of a "wall" we are heading for.

This issue of God's "existence" was the pretext of intensive discussions for years but this was exasperated by the recent progresses in our knowledge of the surrounding world. On this basis the English biologist Richard Dawkins[13] claimed that "God is no more than a delusion."

---

[9] "A life of its own", Michael Specter, *The New Yorker*, September 28, 2009.

[10] "Praying with machine", Stephen R. Garner, *Stimulus*, March 12, 2004.

[11] *Robot: Mere Machine to Transcendent Mind,* Hans Moravec, Oxford University Press, 2000.

[12] *The Future of Man*, Pierre Teilhard de Chardin, Harper and Row, 1964.

[13] *The God Delusion*, Richard Dawkins, Bantam Books, 2006.

His book was soon a bestseller but he was actively contradicted by John Lennox[14] (an Oxford mathematician). God only knows what's what!

### Trans or Posthumanism?

Transhumanism could emerge one day, but likely not explosively. This is like the man on the moon: some went there, but not everybody; same thing for the "improved man" — the first to dare take the leap will be an "explorer" who likely will be holding the bag! There will certainly be several versions which will successively benefit from the improvements.

As for the moon conquest, the budget will be rather high and the results unpredictable. We are not there yet in the confrontation between two different "humanities." There will also be a model of trans-society to be invented. Would it be more rational? Would it tolerate the fantasy and aberrations of the current ones? Would it be sensible to make an "improved soccer player"? Would human reproduction be regulated, calibrated, selected, and optimized? What to do with primitive, uneducated peoples; would they have to be authoritatively retrained?

# The Improved Man

Transhumanism promises that man can be repaired, modified, improved, or even "enhanced."[15] Two aims are being pursued: longevity and brain enhancement.

### Longevity and Brain Sustaining

Extension of life is still ongoing with the progress in medical care, vaccination, prevention, and detection of diseases with improved instrumentation (with the help of computers). New biological means are to be mastered soon with stem cells, genetics, and so on. It is foreseeable that cancer, diabetes, atherosclerosis, and even Alzheimer's will eventually be defeated, as smallpox, poliomyelitis, and even tuberculosis were in

---

[14] *God's Undertaker: Has Science Buried God?*, John C Lennox, Lion Books, 2007.
[15] See the movie *Iron Man* (2008).

the past. That will certainly contribute directly to substantially lengthening the mean life.

Already, mean longevity (in France) is established at 81 for women and 79 for men but it is also currently established[16] that if the statistic is restricted to people who take care of themselves or are wealthy enough (in the U.S.), the peak of the curve is shifted to some 12 years more for both men and women!

More speculative and disturbing "progress" could be expected from the genetic selection of embryos, which is already in place to eliminate trisomy 21 but is worth being ineluctably extended. This is a direct way to artificially reconnect with Darwin's natural selection!

In the same domain of genome engineering, it is worth mentioning a very recent and promising technology called CRISPR-Cas9,[17] which could become an awesome tool. This enzyme has been shown to allow cutting the DNA double helix to make it possible to induce changes in the genome of human embryos[18] (as stated by Junjiu Huang in Canton).

## Brain Enhancement

The first idea would be to use natural genetic selection to improve brain specificities, as is already done for animals to improve a particular specification. This process would certainly work but would require waiting generation after generation, which is particularly lengthy. We are burned by a sense of urgency to defeat Moore's law.

Alternatively, it could be faster to use *in vitro* fecundation: stem cells from individuals selected to generate an infinite number of gametes which could be sequenced, sorted, and fertilized, giving rise to a new generation providing improved new stem cells and so on. The process, then, would be significantly faster. But to get what? This is a blind process; as long as adults are not involved, we would know nothing about the results.

---

[16] *Longevity in the 2.0 World: Would Centenarians Become Commonplace?*, Jean-Pierrre Fillard, World Scientific, 2019.

[17] CRISPR stands for "clustered regularly interspaced short palindromic repeats."

[18] A brief overview of the issue can be found at https://en.wikipedia.org/wiki/CRISPR_gene_editing

That could lead to pure monsters. The epigenetics is not yet sufficiently developed. Minor but uncontrolled changes to our brains could generate unexpected improved performances as well as degradations.

All of that with the aim of obtaining an intelligence for which we don't know precisely which direction to take: philosophy, mathematics, art creativity, technology? Does the intelligence of the street sweeper need to be enhanced? We don't need only intelligent people in our societies (see our leaders!). In a purely eugenic vision, we could better look for obedient, compliant, dependent people.

However, genes are not everything in the establishment of IQ; the shared experience is also important in the building of a brain.

Would we have to synthesize humans only in order for them to be competitive with the machine? Would it be necessary to get copies (by cloning)? Some countries, which gained a technical edge and have not any Christian bias, could turn to a programmed eugenism.

How to selectively boost a creative, inventive, artistic mind? Would it be thinkable biologically, genetically? Epigenetics and neuropharmacology are still in their infancy; more data are required to reach conclusions. Would doping the hypothalamus be a solution to boost the neuron production?

## Toward a Higher Intelligence?

What criteria for the limits? Do we take ourselves to be God?

### Nick Bostrom Emphasizes a World Domination by Machines

This scenario seems quite unrealistic (at the moment) at a large scale, because this would imply that robots would constitute a global society able to fulfill on its own any of its needs. This is certainly not for tomorrow. Would this society be selectively technical, military, business? Would it be unique or multiple or competitive? Would the frontrunner be favored? There are so many questions to be answered.

Comparatively, the current human society, even though far from perfect, still enjoys advantages which are hard for a machine to compete against — essentially the flexibility, the adaptability to any new situation, the fantastic diversity of the competences (even at a very basic level), the creativity and imagination to find solutions. To date, it is humans who make the robots and put them into action. Even if equipped with an effective AI in specific domains, one can hardly imagine a fully autonomous robot society.

## Do We Take Ourselves for God?

Would a superior AI (internal or external) allow us to fully dominate nature? That seems quite crazy and, once again, "knowing" is not "having the power to do."

We are definitely not God, even though "playing God" has long been available in some games of life online. John Conway was a mathematician, a charismatic pioneer in this theory of games. But let us be serious — God is not a game.

Since the primitive ages, as soon as he got down from his tree, man understood he was in a dangerous world which results in death for everybody. These threats led him, instinctively and everywhere, to imagine a psychological help to survive these difficulties: he invented gods, which later became God. Of course, each tribe invented its own gods, and also later its own God when it got larger. There is a large diversity of religions in the world, but they are all imprinted in the depth of our minds from infancy (even unconsciously for people who claim they are atheists); part of our soul, like the coat of Nessus. Be that as it may, the basic metaphysical questions "Who am I?" and "Why do I exist?" remain unanswered, even with the help of a supercomputer.

Religions are pure creations of the human brain in the quest for explanations of our destinies. Some are peaceful; others claim affiliation to the Devil.

They all lead to wars but remain necessary for collective and cohesive equilibrium of peoples. When religions disappear, chaos prevails. Then the question is: would religions (and which one?) have to be taught to our future computer equivalent?

At all times science has been accompanied by a falling away of the traditional certainties and has generated fears for the future. God is a haven where we have all been deported.

These religions have been so deeply involved in the subconscious, as an unwavering conviction, that they have given rise, for millennia until today, to bloody wars with, on each side, the promise of paradise. Yet many religions (some do not care about this point) command: "Thou shalt not kill." This mandatory order has been intentionally taken by Turing as the first rule for robots! And this rule is still, in the same way, confronted by military! Nothing new to the robots! How do we make them kill only the enemies? During World War II, fighter planes were equipped with an Identification Friend or Foe (IFF) beacon in order not to be shot down by friendly anti-aircraft defense or fighters.

Currently, however, the immediate dangers have turned smaller (I hope it is so) but they still are there, with the corresponding mysteries, in our modern lives and, anyway, death always remains there as the ultimate threat. So would a robot, a cyborg, a transhuman-computer-fitted man, still require gods or God? If the intent is to authentically copy a brain in a computer, the answer is obviously "Yes."

Would it be an improvement for the emulated brain to be freed from the notion of religion? That is a matter for debate, but the resulting AI would no longer be humanlike without an (possibly) artificial god.

How do we teach religion to a robot? Should it be mono-religious or is religion optional? If a robot is as intelligent and humanlike as to acquire religious feelings, should it then not be considered fully human, and what about its civilian rights and duties (remember that a motion has already been put forward at the European Parliament to take robots as "electronic persons"!)? The robot may have the upper hand over the human being to the extent that it is not afraid of death or injuries or suffering (in case of any accident it is only to be repaired and start again). So, would religion be of any interest to it? Even HAL was not afraid of being plugged out!

As a matter of fact, some religionists, like the Mormons or the Buddhists, consider their religion compatible with transhumanism because we are only inspired by the will of God, which guides our destinies.

## Is the Computer Able to Create Intelligence?

If intelligence is limited to finding a solution to a particular, well-defined problem, I would say yes. It could be taught to do the job; but if a human-like or higher intelligence is required, I guess this would be presently unthinkable.

## What to do with Humanity if the Brain is Overcome by AI?

Some say that robots or cyborgs or any of these super-intelligent machines would rapidly get so much power in the societies that they would take decisions in place of the humans, who would then be reduced to a sort of accepted slavery because of their intellectual inadaptability to the "robot society."

In fact, this splitting of a society between culturally adapted humans and unproductive, uneducated populations already exists and is increasing with the current uncontrolled and excessive rise in the world population. However, presently, a society needs simple workers as well as Nobel Prize winners. Would that have to be changed with robots?

Obviously, if "improved men" are to become a reality, they certainly would constitute a particular category — but what exactly to do in order to fill high competence jobs? Is that a real ambition in life? Maybe to some people. And, to push the idea a bit further, would ordinary men still be necessary and to what extent? We have unquestionably entered a eugenist paradigm.

Presently, it is well admitted that man has to keep his hands on the final command of every robot and has to be able to make the right decision if something goes wrong. But would this point of view have a future?

The recent crash in France of a plane with a German pilot who committed suicide (killing all the passengers and crew) triggered the idea that, after all, man is not as reliable as expected and should be better supervised ... by a convenient, clever, fast, and rational robot. Aircraft industries have already proposed a panoply of solutions (NASA: Auto-GCAS) for which, in case of something happening out of the normality (but how to define "normality"?), the robot (possibly helped from outside) takes full control of the plane,[19] independently of any local human influence.

As regards the preceding plane domain, there is, however, the counter example of the Habsheim crash, where the pilot in a demonstration flight

---

[19] Philippe Fontaine, 01net, No. 819 (2015).

at low altitude, pulled back on the control column in an attempt to steer clear of the unforeseen approaching trees. Alas, the computer did not obey, convinced it was in a landing operation, and kept the command down until the crash, in spite of the efforts of the pilots.

There is a very preliminary step to give a superior power of decision to robots in many circumstances and contexts (nuclear power stations, chirurgic operations, trade markets, credit card supervision, and so on). Last time (May 6, 2010) such a robotic process was used in the U.S. equity markets, a large seller initiated a dedicated sell algorithm and it happened that the system diverged, thus triggering panic among the traders and leading to absurd decisions. This resulted in a downward spiral and in the loss of trillions of dollars. It was hoped that the robot would finally obey the programmed stop order — too late, unfortunately, the damage had already been done.

This is to say that robots have to be taught in detail what they have to do in any imaginable circumstances. Otherwise, as has been demonstrated, these stubborn machines will do as they want, leading to catastrophes.

## The Global Brain at a World Scale

Distributed cognition, extended mind, hybrid self, hybrid assisted limb (HAL) — all these terms flourish in the literature, but they are becoming realities.

### Homo-Googlus and the Hackers

Vernor Vinge was a mathematician and software science specialist who wrote novels in the science fiction domain; he was famous for imagining the Singularity and a society of "advanced hum" for the first time.[20]

Now, computers have made progress and we know that dangers could arise from hackers; brain emulations could, as well as any computer assembly, be hacked with digital viruses or malware without the need to destroy the hardware, or the software, which could be slightly modified to make the attacker benefit from computing power for his own use.

---

[20] *The Witling*, Vernor Vinge, Daw, 1976.

In that way assassination becomes virtual and much easier than for the original human! Hopefully, for the moment, the rights of the emulation's personhood do not present any priority.

The idea consisting in the gathering of humans to achieve a collective intelligence is certainly not new; every industrial enterprise, every organized society follows this practice with success (and corresponding flaws). However, this starting idea could certainly be optimized and resized[21] to a larger scale using the Internet's global dimension. We are now familiar with the Internet, but this medium remains underexploited; further improvements can be expected and are surely to come soon.

## Brain Hacking

How do we hack a brain?[22] Two ways could be proposed:

- The first could be to read into the brain to investigate what is good or bad, what is worth copying or modifying. This assuredly is on the way at GAFA[23] under different approaches.
- The second way is to write inside, using the Nanotechnology, Biotechnology, Information technology and Cognitive science (NBIC) technology under development.

However, there are also softer and well-known methods which have been commonly used for a long time. Education and learning as well as psychiatry are in some way types of hacking. Now, neuro-hacking is on track to get an industrial shape. Google, through its program Calico, already experiments with neuro-manipulation in order to keep the brain in good shape for pushing the limits of death further. As Sergei Brin put it: "We are going to make machines which are to reason better than humans."

---

[21] *A Fire Upon the Deep*, Vernor Vinge, Tor Science Fiction, 1993; "The coming technological singularity: How to survive in the posthuman era", Vernor Vinge, NASA Lewis Research Center, Vision 21: Interdisciplinary Science and Engineering in the Era of Cyberspace, 1993.

[22] *Conference Les Assises de la Sécurité et des systèmes d'information*, Laurent Alexandre, Monaco, October 14, 2014.

[23] Google, Apple, Facebook, Amazon.

Google claims that such a machine will be achieved which can answer our questions before we are able to phrase them. Information located in our brain is already reachable using electromagnetic techniques, eye movements, medical files, the Web, prostheses, or DNA analyses, all this information being cross-checked and assembled. The same curiosity is developed at the NSA,[24] which wants to stand out through overseeing everything, even at the individual scale.

In animals it has become possible to modify the memory by means of transcranial stimulations or optogenetics, whereas in humans "hybrid thinking" has begun to be explored in relation to a machine (implants or external). The Chinese are sequencing the DNA molecules of gifted individuals to decipher the secrets of their brain behavior.

Our senses can be easily fooled; experimentally introducing a decoherence between the image and the sound can generate confusion over the meaning of the words and lead to "sensorial hacking."

In any case, what could be worth hacking in a brain if, as Michael Abrash put it, "reality is an illusion; reality is not what exists but what your brain records"?

All of this demonstrates a growing curiosity which, as an unavoidable consequence, will appeal to the hackers and make it necessary to implement "neuro-security" for our brains.

The domain of the hackers is constantly increasing; it is no longer restricted to computers but extends now to any application: recently it appeared possible to take control of a "Google car" from the roadside and make it to obey a hacker in place of a potential driver!

## The Cloud, a Distributed Intelligence

Google (and Baidu, its equivalent in China) corresponds to a utilitarian form of intelligence which is still imperfect and limited in its means and aims. But we have to consider that it is only 15 years old. Search engines will soon be improved so as not to be limited by keywords; the time will come when we will speak to Google in sentences and it will answer continually and with clear competence. It is now its turn to face the Turing tests, in the opposite direction.

---

[24] National Security Agency.

The day Google intelligently "understands", everything will be upset, and we will directly reach Mr. Know-It-All. Google will then have to master language (AI linguistics) in both directions; this is on the way to being done, at least in a sense, with search engines which spy on telephone conversations and analyze them to find patterns. The National Security Agency does the same, and stores information to track down the suspects (and maybe others) and assess (that is to say, understand) the collected information.

This does not necessarily require copying a human brain but can be developed on different, and maybe more effective, research structures.

The common fear is that such a global and advanced organization could lead to just-as-global servitude and dependence for humans.

But it is not necessary to wait for a robot dictatorship; our servitude has already been openly achieved by the big companies, namely Samsung, Google, IBM, and many others, in a watered-down but psychologically intrusive form: the "entrepreneurial spirit."

Under the cover of preserving the welfare of their employees (that is to say, improving their productivity), these companies maintain a close and invasive "cocooning" of the individual and even his family: meals, sport facilities, leisure centers, games (all of that for free), even housing and dedicated supplies (Samsung has created a dedicated "Samsung town," where people can do shopping and find everything they need and, by the way, are discouraged from going outside; this is really a deliberately accepted "concentration camp").[25] All that raises the interest of the employees and sets the "entrepreneurial spirit" in accordance with a pre-established mind model.

That constitutes a true and soft invasion of the minds, which are unconsciously mobilized by only the interests of the company and in some way forsake their individuality to become consenting and formatted slaves, concerned only with contributing to the prosperity of the company, which, as a counterpart, generously takes care of them. This is an accepted win–win agreement, but also terrifying mental conditioning.

The only thing that remains will be to fine-tune the AI with improved means of accessing the brain in order to make possible an extension of the

---

[25] That reminds me of an old British TV series with Patrick McGoohan in the role of "Number 6": *The Prisoner*.

method to the whole human activity, and we should become happy "human robots."

## Global Brain Institute (GBI)

The Global Brain is a metaphor for the Internet which connects all humans and their artifacts. This network becomes more and more intelligent, due to the wealth of information it stores. Then it starts to resemble a real brain at the planet scale. The connections and pathways of the Internet could be seen as the pathways of neurons and synapses in a global brain. This "planarization of humanity" was considered an irreversible process by Teilhard de Chardin as early as 1945!

The Global Brain Institute (GBI) was created in 2012 at the Vrije Universiteit Brussel to improve research on this phenomenon. The aim is to anticipate both its promise and its perils with a view on getting the best possible outcome for humanity. The aim could also be to develop other dimensions of the brain that are, today, masked by a hostile world and still arranged for survival.

The question[26] now is: is the Internet evolving into a global brain? The Internet is already a sphere of collective consciousness. Julian Huxley prophetically stated[27] that "we should consider inter-thinking as a new type of organism whose destiny is to realize new possibilities for evolving life on this planet." But, with such a delocalization, psychiatrists warn that children currently find it more gratifying to play in a virtual world and this could lead to uncontrolled violence and depersonalization.

## The Energy

As already mentioned, knowing and being able to do are not the same. Intelligence and power are not the same and belong to different needs in society. We really should not mix things up.

---

[26] See for instance: http://www.huffingtonpost.com/cadell-last/internet-global-brain_b_3951770.html

[27] "Introduction", by Julian Huxley in *The Phenomenon of Man*, Pierre Teilhard de Chardin, Harper and Row, 1959.

When we are making a comparison between brain and computer, there is one small but essential domain which ought not to be forgotten (it often is): energy.[28] Indeed, in this domain, each of these two supports correspond to drastically different needs and sources. The brain stays in a purely biological domain, whereas the computer only feeds on electricity.

The brain finds its energy from the oxygen carried in the blood through hemoglobin molecules. This oxygen comes from the lungs via breathing while the heart that pumps the blood is further driven by the alchemy of food. The brain is the organ which uses by far more oxygen (energy) — more than the heart or the other muscles. The mean power dissipated in a body at rest lies in the range of 0.1 kW, the most important part being devoted to the brain, night and day. An interruption in supplying blood to the brain can be deadly in the short term (stroke).

However, brains are numerous and all different from one another; each man mandatorily has one; it cannot be dissociated from the corresponding body, and its size as well as its energy consumption is roughly the same from one person to another. One does not yet know how to make an isolated brain survive in a vat, or to graft or clone a brain.

In the case where one brings an external complement to a brain (implant, biological addition, or interface), the problem will immediately arise of bringing it its own energy. This will hold as long as we are not able to make a biological entity directly compatible with the blood flux. But we are not yet there, even though some components have already been completed.

Concerning computers, the size and energy consumption vary to a very large extent; from the isolated chips to the big data center there is a huge diversity. Current robots all encounter a tremendous limitation on their autonomous life in the source of energy. Those which are trying to imitate humans or animals in the execution of a task, such as Big Dog, must spend a great deal of energy to merrily carry a heavy battery with them. The more they make efforts, the less freely they can operate, and when the battery is empty the robot becomes a simple material object without a brain and without "life." However, unlike the human equivalent

---

[28] *Our Final Invention: Artificial Intelligence and the End of the Human Era*, James Barrat, Thomas Dunne Books, 2013.

(namely concerning the brain), the robot is not dead — it just needs recharging of the battery to bring back strength and intelligence unaffected; this is not the case for a human if he stays too long without food or air!

The Internet is an especially huge consumer of energy in steady growth. The work capacity of these centers is directly related to the many hard disks involved, and obviously their number increases constantly. The computing power of these centers will ineluctably be limited in the near future by the electrical power required and the corresponding costs, even if the specifications of the memories are constantly improving.

In the U.S., the big data centers are mostly installed close to an electrical plant in order to reduce losses during transport; in Virginia, abandoned coal mines have been reopened, thus revitalizing a lost business but also revitalizing pollution. How long will we be able to tap the natural resources with an ever-rising world population?

In France, where free space is more limited, data centers are often located at old, disused industrial sites, close to the downstream users — that is to say, inside the big towns, which induces complex connection problems with the grid to get enough power.

In the competition between the computer and the brain, the availability of the energy sources will certainly become a crucial technological issue. The Internet cannot grow indefinitely; all our energy resources cannot be only dedicated to feeding its hard disks, whatever the technical advances that are to occur. The only solution in the future (if any) could rely on the biological computer combining the sobriety of the biological cell with the efficiency of the transistor with the aim of obtaining better compliance with the brain. But a biological Internet is certainly not for tomorrow and even a way to "feed" it remains to be found.

Renewable energies are non-permanent, unpredictable, and often unavailable precisely when they are needed; in any case, they all require a reliable complementary source to back them up at any moment.

The intelligent computer, whatever its kind (implant, android, cyborg, or connected) will be facing not a single human but assembled societies which it will be itself the product and which brings it its energy. It will depend on them no matter what it can do, as long as it is not able to fully support itself. Besides its possible high intelligence, it will have to really become as efficient and versatile as humans are.

# Chapter 13

# A Way for Brain and Computer to Work Together?

Our current 2.0 civilization is full of unexpected but however foreseeable surprises coming from the biological world. That shows us that Nature (some say God) is not ready to abandon his eternal leadership over the human destiny!

Everything which has been said here must be seen now in a new context. A new field has opened. The name of this global upheaval is: COVID-19!

This virus has taken the global medical scientific community aback; its nature and its behavior has been unexpected and required reconsidering from the bottom up all our previous knowledge about viruses. This outbreak has caught the whole world by surprise and to such an extent that some felt that this new stranger might not belong to a natural origin but more likely to an artificial genetical construction escaped from a demonic biological war laboratory (either voluntarily or accidentally) in Wuhan.

All our current knowledge, means, and 2.0 facilities as computers and Artificial Intelligence are of no help to contribute in finding an immediate and efficient answer to this plague. For the first time we are facing right now an economic and social turmoil in the world and this requires lessons to be drawn from the experience in order to organize the future in case of a repeated similar viral aggression. Our goal is not only to cure the disease but also to prevent it.

Right now we are in the biggest crisis since World War II. At the moment, the computer is absolutely of no direct help and the brain had to be deeply activated to understand the arising new issue. Later, when things will have gotten clearer, no doubt that the computer, taught by the brain, might become the ultimate recourse. It is worth mentioning also that giving large powers to the computer over our destinies will mandatorily require a dedicated tight control of it by the brain.

This deserves an extended examination in this updated version of the book, providing a new appreciation of the relations between Brain and Computer.

## Epidemics: Microbes vs Viruses

Epidemics (epizootics) are in no way a new phenomenon; they are our plight from all eternity; antiquity abounds of famous examples where the ultimate explanation required divine interventions.

Epidemics or even pandemics were not made to last indefinitely, they vanish when the virus weakens over time after too many replications or muting in another downgraded genetic shape. This also occurs when finding host cells becomes too difficult. The extinction remains a natural process but may take a long time to fully complete.

Epidemics may come from a bacterial or as viral origin. In the former case the battle is conducted with medicines and more especially antibiotics; in the latter case these medications are of a very limited use and the only weapon available remains prevention (confinement and masks to refrain the virus expansion) and vaccination. Some virus and bacteria giving rise to epidemics are indicated in the table here above; some are inoffensive or even beneficial for the human, others (most of them) very dangerous.

Microbes and viruses mainly differ in their constitution and especially their size: the former are in the micron range the latter clearly smaller, most of them in the 10 nanometer range and that is why they were so difficult to discover with only optical microscopes.

Antoni van Leeuwenhoek invented in 1676 a kind of microscope which made him able to discover a new living world at the small scale: cells, bacteria, animalcules, and other creatures unsuspected until then

**Table 13.1.**  Possible sources of epidemics

| Virus | Bacteria |
|---|---|
| Spanish flu | Plague |
| Season flu | Cholera |
| Asian influenza | Typhus |
| Poliomyelitis | Diphtheria |
| Smallpox | Tetanus |
| Rabies | Whooping cough |
| Yellow fever | Koch bacillus |
| AIDS — HIV | Streptococcus |
| SARS | Staphylococcus |
| Ebola | Salmonella |
| MERS-Cov | Measles |

(spermatozoids). It was this very primitive means (roughly speaking, a magnifying glass) that gave rise to the more effective optical instruments we are currently familiar with.

Then, throngs of microbiologists began to work and (Brain) brought to light the relationships between these microorganisms and some particular diseases already known. They became very skilled in identifying, cultivating and selecting these living species; they were even able to recognize in the details their metabolisms.

Doing that, they were also able to derive adapted medicines, protective behaviors but also vaccines which allowed preventing the risk of infection and a way to stop it before it spreads. In spite of that, epidemics continued to happen on a regular basis as the notorious pests coming from Asia with the commercial ships of the times.

Already people were fully disarmed against the plague. To get away from it they hid behind gruesome makeshift masks in hope of driving out the curse. Already they claimed for home confinement and collecting the dead that are waiting to be buried. Nothing new under the sun. However they had no smartphones, no TV channels, no Facebook, no Webmail (as we have now) to share the burden among each other. They only had the

**Figure 13.1.**   The crow mask of the old times.

brain to rely on in the absence of any computer. Figure 13.1 is reported a typical costume of the "physician" of the times!

Epidemics are caused by contact of people through breathing, spluttering, dirty hands, having sex, etc. Evidently current population movements, travels, and so on strongly favor the dissemination of the miasma at a world scale.

## What Especially about the Viruses?

Biologists in their experiments had become accustomed to "concentrate" their cultures of microbes using special filters made of "Chinese porcelain" which retains the microbes and lets the solution flow. They quickly discovered that such evacuated solutions might also be infectious even if they no more contain bacteria. These unknown pathogenic germs (toxins) were unobservable but still active. The microscopes of these days were hardly able to resolve micronic objects but no more.

It was not until the 1930s that the first electronic microscopes made it possible to improve the resolution down to the nanometer domain and make the first observations of the viruses. Their shapes (spheres, sticks, or other more complex configurations) and dimensions (some 100 nanometers or less) vary to a large extent.

Their belonging to the living world was and still is a matter of controversies as their closeness with the chemical unanimated molecules makes them not easy to differentiate. But whatever their proximity with the living world viruses do not belong to a particular gender and they even reproduce through specific molecular reactions.

Even though viruses form an integral part of the whole living world; we are living among multitudes of viruses and the oceans, for instance, harbor a huge reservoir of divers and varied viruses which remain associated with cells suspended in the water. Viruses can be commonly found as well in animal or vegetal species, some are harmless while others are virulent.

Then viruses are in no way living cells but more likely infectious particles containing genes confined in a shell. They reproduce from their own genetic material, DNA or RNA, which compose their genome. To reproduce they need to parasite a living cell adapted to their nature, if not they disintegrate. If let in the air they spontaneously vanish within a short while especially when the temperature is high and the air dry. We do know some 5,000 different species of virus and winter is the favorite season for the multiplication of the viruses and so allowing epidemics to develop.

As soon as they house in a host-cell, viruses multiply by diverting the metabolism mechanism of the cell and so generate copies that are expulsed from the cell to live their own life. These newborn cells may be pure copies of the model or, as well, mutated species going to reproduce as they are and that shall make fighting more difficult.

Among the various familiar species of viruses more or less virulent there are: rabies yellow fever, small pox, flu, poliomyelitis, etc.... Most of them are now fully mastered by adapted vaccines. The genetic analysis and flow cytometry allow easily identifying these known species. Other newcomers such as HIV, SARS or MERS require a more delicate analysis. Presently, the main issue with COVID-19 is that no one's immune system has seen this virus before and we are completely unaware of it; then we are all easy preys.

Viruses can be grown in glass or plastic vessels with specific animal or vegetal cell cultures. Their safeties are to be tested first on laboratory animals before carefully passing to humans. With the time going, a huge knowledge has been accumulated (in the brain) on the viruses and the data stored and updated in real time ( on the computer). This is a prime example of what a brain/computer collaboration could be.

The result of that is quite variable: some viruses have been completely eradicated (mainly poliomyelitis, smallpox, diphtheria) by means of preventive vaccination[1] of babies. Some viruses mutate frequently (as the seasonal flu) and require the vaccine to be annually adjusted so the efficiency seldom reaches 100%. Epidemics depend on population movements which favor tcontact between humans over long distances.

The Asian flu was widespread in 1957 and resulted in some 2 million casualties worldwide. The Hong Kong flu also induced 1 million in 1970/71. A special mention merits attention with the "Spanish flu" which has caused some 50 to 100 million deaths in 1918/19 just after World War I and is now also quite forgotten.

Surprisingly these disasters were quickly forgotten; people were more deeply concerned with Apollo missions to the Moon or the Vietnam war. The world never stopped turning and life does go on.

In these not so remote times there were no TV channels, no smartphones, no Facebook and social networks to comment on the current events and spread fear and panic. Could we have become psychologically more fragile or did the 2.0 world make us think that we are now in a riskless environment where conquering death is for tomorrow? Nature has come back as a reminder.

## Previous and Current Viruses' Epidemics

The "natural" evolution of an epidemic, in general, classically follows the same scheme. The phenomenon starts with a fast exponential rise (the virus is strong and the human's immunity system unprepared to respond

---

[1] Let us recall that a vaccination consists in injecting the patient with an attenuated strain of the virus in order to stimulated the immunity system of the body which will then remain on alert.

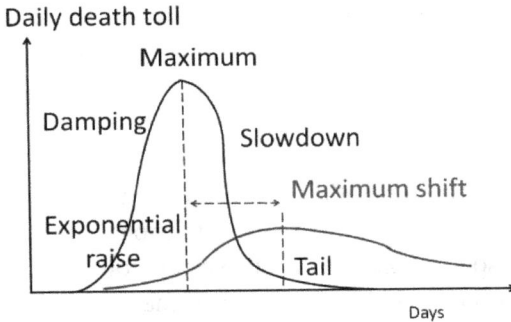

**Figure 13.2.**   Classical evolution of an epidemic.

to the aggression). Then a second phase comes when the virus gets tired (perhaps due to overactive reproduction) and the curve of the daily death toll begins to saturate until reaching a maximum which is followed by a rapid slowdown and the final phase displays a more or less long tail before reaching the end of this transitory phenomenon.

This schematic behavior is observed whenever the virus is left free to evolve and the peak is sharp and high. Instead, if counteractions (i.e., masks and distancing) are introduced to restrict the process, the peak can be strongly attenuated and the maximum shifted toward longer times (see the red curve on the plot). People think they have defeated the virus but, in fact, the integral of these curves often lead to almost the same total amount of casualties and this remains a Pyrrhic victory. The only benefit is that we have a longer while to treat the patients and the number of daily patients in the hospitals is less even if the extension is more sustainable.

Of course, when dealing with a new virus, as it is the case presently with COVID-19, our means of defense are limited and we are close to a situation of a spontaneous evolution in the shape of a sharp peak. We are brought back to the ancestral ignorance before this scourge and the available means always remain the same: confinement, masks and distancing to prevent as far as possible the contact with the germ.

However, biologists are familiar with many families of virus, even those which have been recently discovered such as HIV, SARS, H1N1, Ebola, or Zika. The evolution of the epidemic is similar to those caused by microbes with the difference that usual medicines are rather inefficient.

The only way to counter the virus remains to keep it at a distance and shut down the first cases detected. "The only way forward is to suppress cases and clusters of cases rapidly" said Dr Tom Frieden (centers for disease control and prevention). The only way to win is to know first where the enemy is.

This strategy supposes individual distancing with masks and collective confinement to limit contact. But a tension is to rise into the pandemic: extended social isolation is not so easily tolerated in the long range and the return toward gradual reopening of the society must be carefully managed.

### The Special Case of COVID-19

This particular species of virus belongs to a family of similar coronaviruses (HCoV) which are known to induce breathing difficulties; they often induce minor injuries but may also be more serious with immunosuppressed persons. The annotation "Corona" comes from their viral envelop covered with a crown of glycoproteins as can be seen on an electronic microscope image.

COVID-19 is suspected to be carried by a kind of bat or by pangolins which animals are usually eaten in Asia. This is a newcomer in the family of HCoVs; nothing is known about it and we should learn everything from scratch.

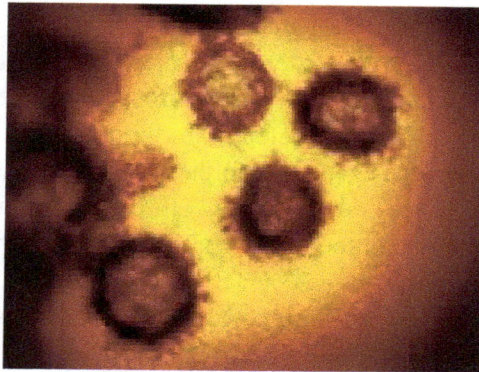

**Figure 13.3.**    COVID-19 image from an e-microscope.

It is rather sad to note that in our 2.0 world the knowledge and the current means are useless. We have come back to the Middle Ages with the only last primitive resort to mask and confinement.

The known symptoms of this virus attack usually are: cough, fever, shortness of breath, or diarrhea. The viral load required to give rise to a transfer depends on the support where it was left and on the delay. In the special case of COVID-19 the contagiousness is high and the virus quite resilient. The delay of hazardousness may vary from some hours to days.

The success of Taiwan in this fight comes from the timely alert early on, travel restrictions, quarantine protocols, widespread body temperature monitoring, disinfection steps, coordination of the production lines of surgical masks.

But everywhere the question remains: how long should the lockdown be maintained and when and how to release it gradually? How to safely get rid of that supposed engineered "bioweapon" released in an "accidental breach"[2] should be classified in two categories:

(1)  Short term strategies:[3] (a) serological testing anti-bodies in the general population (b) develop reliable antigen tests to diagnose those who carry the virus (c) install contact tracing by apps technology to rapidly identify the contacts to quarantine to prevent the spreading.
(2)  Long term strategies to eradicate the virus: drug and vaccine; however, it still remains simply out of hope to get social and economic recovery within a year.

Nevertheless it is established that one does not die from the virus but from the disorder it has generated in the body. The virus behaves like a match, it only lights the fire. When death occurs the virus has already disappeared. That is the reason why a medication such as hydroxychloroquine is of no effect for patients in an advanced process. That is also the reason why young and healthy people are well more resistant than old or crippled people who may suffer from other failures.

---

[2] "US-China tensions cloud virus origins", John Walcott, *Time*, May 8, 2020.

[3] "How to prevent a depression", Klaus Schwab and GuidoVanham, *Time*, April 27, 2020.

## What Can the Brain Bring?

The brain is starting from nothing to find a solution. Some intuitively supposed that some well known medicine could be of some help. For instance Professor Didier Raoult[4] made the buzz in Marseille with the presentation of a possible cure against COVID-19: the hydroxychloroquine, a derivative of the quinine previously largely used against malaria.[5] I absolutely do not know the reason why this therapy should be efficient against this virus but it is cheap and easily accessible, so what the heck? I am not alone to support this idea; President Donald Trump also used it as a preventive care but he was nevertheless infected and hopefully has recovered from an experimental procedure.

Then, this is a typical direct contribution of the brain which does not refer to the computer. Over time, nevertheless, the more efficient way to fight a virus on a large scale remains a vaccine, but that requires a lot of time to be developed and tested before use. Time is an inescapable parameter but in the present case, as pointed out by Anthony Fauci (infectious-disease expert), "[t]his virus is so transmissible, it's not going to disappear".

## What Could the Computer Bring?

So let us turn now toward the computer, and what might be its contribution. Assuredly this pandemic might be so huge that it deeply changes our societies from the bottom up.

The computer tends to be good with huge amounts of data to be stored, crunched, and quickly processed. It is also the favorite support of Artificial Intelligence which limits are yet to be discovered. Deep Learning is perfectly adapted to sort out populations, to exploit medical files. Assuredly the computer would not afford a solution but will certainly allow reaching it.

---

[4] "Dictionary of infectious diseases: diagnostic, epidemiology, geographic distribution, taxonomy, symptomatology", Didier Raoult and Richard Tilton, Elsevier, 1999.

[5] I am personally indebted with this medication that saved me from death's door when I was a young boy. I must testify that it did not induce any side effect but I nevertheless cannot testify if it could be an effective tool against a virus.

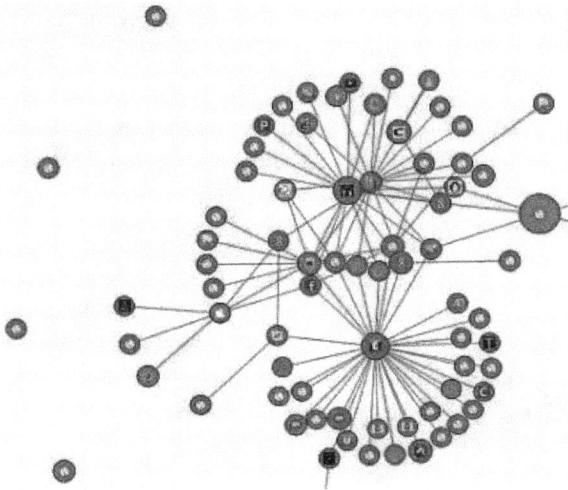

**Figure 13.4.** The way the virus spreads from an initial single contact (big blue circle on the right) to many others.

Already the computer is not afraid of terabytes of information and that will be a tremendous means to absorb the data relative to a large population of humans as it begins to appear in China, but also elsewhere.

Street cameras, drones, satellites already gaze at the people below and record personal data using face recognition. In the present case of tracing a virus it has become possible to track people at distance, identify them, and even take their body temperature thus allowing identifying possible virus carriers and by the way also possible contacts that could also be infected. This is known as a "tree detection" process.

South Korea's contact tracing program even involves GPS tracking of all new positive cases and also imposes wristbands for scofflaws! Some fragrance of dictatorship? But officials were so able to immediately notice a new outbreak[6] and identify its source (1), determine the number of newly infected (102) and a possible cluster of new cases (5,500). They so were able to react rapidly to contain the virus and limit new outbreaks.

The medical domain is especially relevant to such a technology and struggles to fit with its requirements. Medical individual files (detailed and

---

[6] "The risk of reopening", Haley Sweetland Edwards, *Time*, May 25, 2020.

updated) are to be collected and that is not so easily achieved with preserving the confidentiality of the information. In the situation of a serious pandemic these limitations are to be softened to safeguard the general interest. The balance might be quite difficult to find, depending on the local situation.

The main advantage of the computer obviously lies in its amazing ability to gather incredible amounts of data. This is reinforced by the various tools offered by the Artificial Intelligence to crunch through the information in these mountains of data and sort out the relevant ones. Deep Learning also allows introducing an appropriate intelligence adapted to the issue involved.

Alphabet recently announced a partnership with Apple to build contact-tracing software for people's smartphones which could notify them if they have recently encountered someone who tested positive for COVID-19.

This first step detection is essentially based on a temperature check or various external signs of fatigue. The very step to follow is an affordable, easy, cheap, fast, and dependable biological test that can be performed on a large scale to screen a full population in search of those who could be symptomatic or exposed to illness. The first layer of testing is routine screening right in a doctor's office. Dedicated machines are already available such as Abbott ID NOW (13 minutes), Cepheid GeneXpert (30 minutes) or Quidel (a few minutes).

Establishing[7] a viral library of the genetic data from the existing 750,000 viruses would also dramatically decrease the time to develop a new vaccine. In a resilient future that will change our preparedness: a digital immunity passport was not imaginable a decade ago that would attest the individual's immunity to a virus they have been vaccinated against. Software developments now are deliberately oriented toward modeling epidemics.[8]

Imagine this epidemic had occurred 20 years ago in a world with limited Internet, no remote work systems, no e-commerce and grocery delivery; a world ill-equipped to detect an outbreak and sequence a virus in days and scale diagnostics in weeks and vaccine in months.

---

[7] "A tale of two futures", Nathan Wolfe, *Time*, April 27, 2020.
[8] "Introducing the outbreak threshold in epidemiology", Matthew Hartfield and Samuel Alizon, *PLOS Pathog*, 9(6), 2013.

Contact tracers have to inform each of the possibly exposed and instruct them to quarantine for 14 days. Smartphones are to play a decisive role in the effort; Google and Apple are collaborating in developing proximity sensors using smartphones and search data; of course strong concerns about the individual's privacy remains to be protected.

Public health workers must identify and inform every person who could have been contaminated. This is a huge operation in which the computer could play a decisive role.

The better defense against a virus is immunity as it can be obtained by vaccination. Smallpox was eradicated thanks to immunization.

It is still unclear if people recovering from Covid 19 can be considered as immunized. Nevertheless researchers around the planet are racing to develop a vaccine but that will likely take at least a full year.

## The Chinese Model of Crowd Controlling

China is by far the most advanced country in automatically mass controlling the population through an outdoor-ready camera network. Face recognition technology currently allows pinpointing anybody in an urban crowd and that has become compulsory in most of the towns. That is poetically called "the celestial eye". There is no limit to the incredible extension of such a "Big Brother" collective means in the control of individuals.

All around the clock citizens can be verbalized in the event of law infringement. Any country would be interested in such a technology in case of a sanitary or terrorist threat and this is the case with COVID-19 and particularly with the new gadget of the thermal camera able to get remote information of the body temperature of any individual identified within a single image.

Then the most reliable strategy against the plague is to build a system to test anyone who is infected or might be infected. That means tracking the people who might have been in contact with infected ones. That requires a widespread testing and huge data collection participating in a more global database of biometric personal data.

That also supposes a highly coordinated system that would track and redirect corresponding supplies to where they are needed. That means a whole material organization to be implemented and computer is the only possible way to achieve the goal.

**Figure 13.5.**   Face recognition is everywhere in China.

An example of detailed automatic identification practiced in the U.S.

However China is not alone in this process; similar efforts are being made in Russia (Moscow: NtechLab), England (London: Centre for Global Infectious Disease Analysis), Israel (Any Vision), Japan (Earth Eyes Corp), even France (CNIL[9]).

---

[9] Commission Nationale de l'informatique et des libertés.

Current softwares are now able to make a perfect face recognition that they are not confused by a changing coiffure or beard, a pair of spectacles or even a mask! Deep Learning allows to unambiguously (?) determining the sex of the person and even its current temperature!

## Current Damages at the World Scale

We are likely in for a long fight with COVID-19 until we get a vaccine or even after. Meanwhile damages accumulate in our unprepared developed societies: slowing activity, intensive nursing cares, vulnerable economies. Even China is strongly affected: as Derek Scissor[10] puts it "it is a two-side coin", either an opportunity to build up domestic enterprise, or a full collapse of its export-reliant economy; or even both! Donald Trump for his own part considers COVID-19 as the "worst attack" ever on the U.S., in view of the fact that (among other things) "China produces 97% of America's antibiotics" whereas "Apple produces a vast majority of its wares in China".

The present and future consequences of the crisis, the economic disincentives, the global policy, everything has turned upside down. The brain brings ideas and knowledge and is compelled to call upon the computer (which remains a tool) to organize the future.

The current facilities of disseminating information in real time are accountable for the fame of this epidemic and the panic that resulted on an individual as well as collective scale.

The United Nations forecast that potential effects in the long range over the global economy as well as that of each country will be disastrous. The International Monetary Fund announced that the world entered a major recession requiring a large scale and multilateral response in the range of at least 10% of the world GDP.

That does not make things comfortable for the brain and the computer has to demonstrate its capabilities.

---

[10] Chinese economy specialist at the American Enterprise Institute.

# Conclusion

All over this book, we have been tentatively trying to discuss the "biggest existential threat" of our times. In the disruptive 21st century, the competition of brain vs. computer is a critical element of a larger technical as well as human revolution which has begun.

In the light of the most recent developments related to COVID-19 it appears that, after a while where both the brain and computer were taken by surprise, it could be expected that they would eventually find a common ground and collaborate efficiently in order to provide us with a global full protection against viruses. However the shadow of Big Brother is still hanging over all our heads as suggested by the Chinese example.

## Who is Afraid of the Big Bad Computer?

The constant progresses of AI and of the power of computers have caused deep concern for many people, who claim that an immediate danger looms. Would the famous "three little pigs" (Musk, Gates, and Hawking) be right to sing in a single voice: "Who is afraid of the big bad computer?" Would they be right to sound the alarm? Would we be able to change the course of our destiny? Are there only bad things to be expected (as stated by Wozniak:[1] "The future of AI is scary and very bad for people... computers are going to take over from humans, no question.")?

---

[1] Apple co-founder Steve Wozniak speaking to *The Australian Financial Review* on March 23, 2015.

Of course, this book does not, in any case, pretend to be an answer to these burning issues, maybe only to assemble comments to shed some light on a subject that concerns all of us and that is changing every day. All of the possible previsions can only be a step toward science fiction that is now so close to reality and we have been very careful not to cross the red line.

At first, it was philosophy that incorporated the sciences (as principles). It was a purely intellectual game. Then came the metrological observation; we are no longer talking in a philosophical vacuum; we get to constructive decisions; the feedback on hypotheses can be physically and quantitatively checked; we are no longer in a sophism but have entered rationality — that is to say, the world of the computer.

There, we have to be careful in our conclusions when dealing with the mind. The human brain remains the only model of intelligence available to us; yet we have to understand "what is inside," as Intel puts it, before trying to mimic or transpose! On the other hand, why not conceive, from scratch, an advanced intelligence based on new criteria and without the aim of copying humans? That is the way planes were invented, no account being taken of the flapping wings model of birds. But, in such a sensitive domain, would we be challenging God, who would be at our very origin?

Consciousness, with the help of the information given by the body, analyzes the data and makes a model of the external world. That is weighted by the subconscious and results in the corresponding actions. But the brain is a fantastic neuronal machine which will not be so easily challenged, even by sophisticated technologies.

## The Invasive Computer

The computer is a very recent invention but its developments, in such a limited time, defy imagination to the point where some believe it could begin to successfully compete with the human brain. It has attained the ability to resolve many issues (even very complex ones, out of the reach of a human brain) as long as they are previously well identified and enough data is gathered.

The computer has indisputably already gained an important place in our daily lives; to such an extent that living without this machine has now become unconceivable. One even unconsciously asks the following question: How did we ever manage before? Would it become so invasive that we could not live without it, and would it be able to invade our brain? Would it, in the end, be able to replace our brain?

These are quite tricky questions, because the brain has strong arguments in its defense and Artificial Intelligence will have still more requirements before challenging the versatility, the creativity, and the subtlety of the brain. However, our collective dependence on the computer will even improve, thus resulting in corresponding fragility and in exposure to new risks. What might happen when this dependence becomes individual, when our brains are "shared"? What could be the limits of our submission?

## The Global Threat

A defense of the brain can also be found from the fact that the brain is not unique but largely diversified, multiple, and evolutive; that is what we call a society, and that considerably complicates the possibility of getting a reliable digital copy.

However, things are not so simple. Aside from the usual shape of the little or the big computers we are accustomed to, there is also a collective, global aspect to be considered which resembles a reflection of our society, which is also distributed all over the world in an enormous cobweb: this is the Web, which already impinges on science fiction.

An enormous database has been assembled which "digests" information. This global base was organized by humans and is called the Web (or, more specifically, the Internet). It does not need the five senses of the body — it is already provided directly with all possible information; it would be able to carry on with that and take decisions. This global computer could certainly be provided with a kind of "consciousness," different from the human one; it could have its own (rational) intelligence and get "educated" through a learning process. It would only need to be taught the rules of appreciation: what is to be considered good or bad, important or accessory, negotiable or definitive.... It also (and this is far from obvious) would have to give a clear meaning of the words it would use to "think" (like a human). In addition, it would also have to draw up a "moral" of evaluation and conclusions before getting into action.

This kind of "digester" could be extrapolated from the human model when we fully understand how the biological machine works. It already has an embryonic nervous system to fuel its consciousness: search engines. All of that inevitably brings us back to "Google, master of the world." Undoubtedly, in a first step, it appears that the simplest way to do it is to use the "humanware" to do the job instead of an army of bots less flexible in use!

Such a digester, set up on the Internet (or an equivalent), would necessarily be delocalized, as consciousness is in the brain. The only limitation on such a network would be the energy required, which cannot possibly be infinite.

A possible rival entity of such a machine could be a "collective brain" made up of all the human brains assembled and supported in a global network, each brain playing a role similar to that of the neuron as is beginning to happen in social networks. This is already emerging. These multiple and random relationships could proceed as the spontaneous "flash mobs," which work a bit similarly to the neuron flashovers which are recorded in the brain before an idea emerges in the consciousness.

That would lead to a competition not between a man and a cyborg but between two collective and global virtual entities. The "ring" of the match would be the Internet or the like.

Would man be doomed to be replaced if it appears he is not efficient enough? The debate is now open, but so many things remain to be achieved. The match is not over — it has just begun, and we are the spectators as well as the actors.

But, please, we are not to take ourselves to be God!

## The Last, but Assuredly, not the Least Comment

Our current 2.0 civilization is full of unexpected but foreseeable surprises coming from the biological world. That shows us that Nature (some say God) is not ready to abandon his eternal leadership over the human destiny!

Everything which has been said here must be seen now in a new context that is COVID-19.

The virus has upset all our accumulated knowledge, all our technical means, all our intelligence (human as well as artificial!). All that paraphernalia is actually of no help in front of the plague. Brain and Computer are bounced back to back, no winner! Now we start again from square one, the time of the seven plagues of Egypt! We are almost powerless.

At the moment we are fully disarmed and the only weapon we are able to propose is "confinement" at home and wear suitable mask as in the old days of the "black death" as shown in the picture below! The pandemic evolves as it wants, following its own natural rhythm, whatever humans do, through a peak shaped curve, as usually predicted by the Gaussian statistic and its classical bell shaped curve; no one can change that. The end will come when the virus will spontaneously be exhausted after it harvested its casualties.

Now Brain and Computer have to work together, hand in hand, in order to come up with a solution for a future which should assuredly repeat again. The problem is not new but the virus is, and a realistic solution must be found now.

The currently inescapable problem should be in the spheres of skills of the computer provided that the brain gives it the way to proceed. Gathering data, classifying the information, crunching the data are usual tasks the computer is performing well. Then the brain should learn from it and set the strategies.

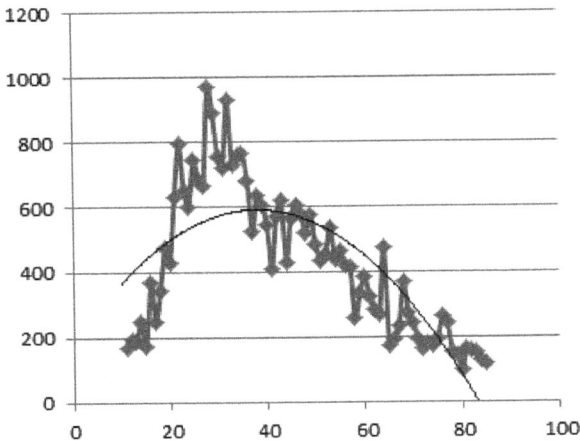

The peak of the daily death toll in Italy.

Such a global challenge implies a juicy market even to elaborate an effective vaccine or to stem future outbreaks. Trustworthy institutions such as GOFAM are already on the track of new pathways. Governments have turned to Alphabet, a parent company of Google and YouTube, for help.

All of that makes a new field for a better relationship between Brain and Computer, which should bring a new hope for humanity, provided that Big Brother is kept at bay.